D1367983

Puzzling Adventures

Puzzling Adventures

TALES OF STRATEGY, LOGIC, AND MATHEMATICAL SKILL

DENNIS E. SHASHA

W. W. Norton & Company • New York • London

For information about permission to reproduce selections from this book,
write to Permissions, W. W. Norton & Company, Inc., 500 Fifth Avenue,
New York, NY 10110

Illustrations on pages 13, 21, 42, 46, 71, 111, 124, 131, and 165
by Gary Zamchick. Illustration on page 107 by Mike Wood.
All other illustrations by John McAusland
Manufacturing by The Haddon Craftsmen, Inc.
Book design by Mary McDonnell
Production manager: Anna Oler

Library of Congress Cataloging-in-Publication Data

Shasha, Dennis Elliott.
Puzzling adventures : tales of strategy, logic, and mathematical skill /
by Dennis E. Shasha.—1st ed.
p. cm.
ISBN 0-393-32663-2 (pbk.)
1. Mathematical recreations. I. Title.
QA95.S473 2005
793.74—dc22
2004021555

W. W. Norton & Company, Inc.
500 Fifth Avenue, New York, N.Y. 10110
www.wwnorton.com

W. W. Norton & Company Ltd.
Castle House, 75/76 Wells Street, London W1T 3QT

1 2 3 4 5 6 7 8 9 0

To my extended Family, loving, energetic, and confidently odd.

CONTENTS

SCIENCE AND FORM

COMBINATORICS

ECCO: A PUBLIC ENIGMA

You'd think that if anyone knew Dr. Jacob Ecco well it would be me. We've appeared in photographs together. We've taken trips together. We've even been kidnapped together. But, in spite of the years I've known him, he continues to baffle me.

The mystery begins with small eccentricities. He wears shorts year-round, even when it's bitter cold. He refuses to wear a tie, even at events in his honor. He doesn't drink alcohol or coffee, but seldom refuses a milk shake. He climbs rocks, particularly enjoying over-hangs ascending from the water, but avoids high diving boards. Tycoons pay him large fees, but his lifestyle never changes and checks lie uncashed in the mess of his desk.

Most detectives rely on observation, knowledge of human nature, and varying amounts of muscle. Ecco relies only on deduction and mathematics. The evidence for success would come from the calling cards of his clients—if he bothered to keep them.

Every one of the cases in this book is accompanied by a small passage in code. Ecco received each one on a postcard around the same time he encountered each case. Ecco has only partly decoded these. The hints seem to lead somewhere beyond the book.

The puzzles themselves came to Ecco in plain text. I present them to you in exactly the form he heard them. When some parts of the cases remain unsolved, you can try to solve them yourself, per-haps inventing new mathematics on the way. This book is not for the faint of mind.

Understanding the puzzles, however, seldom requires formal training beyond junior high school mathematics. Younger kids with sufficient imagination can solve many of them. Even adults have a chance.

Reach, touch, then grasp.

New York City
Spring 2004

P.S. Many of the puzzles themselves have appeared in abridged form in the "Puzzling Adventures" column of *Scientific American*. I expand them here to the full versions that I have heard from Dr. Ecco, often including unsolved parts. Ever the minimalist, Ecco has instructed me to remove all traces of his (or my or even his niece Liane's) presence from those puzzles. He has, however, allowed the accounts previously seen in *Dr. Dobb's Journal* to appear in full. Our little trio has lately acquired Liane's younger brother as a fledgling member, but those adventures are still to come.

ACKNOWLEDGMENTS

I began reading *Scientific American* thanks to the puzzles of Martin Gardner. It was my dream then to write the puzzle column. Dreams are sometimes realized. A. K. Dewdney, former puzzle columnist of *Scientific American*, has helped me in so many ways: he allowed me to write a challenge to readers in *Scientific American* in 1988, invited me to write a column in his excellent zine *Algorithmics* shortly thereafter, and gave me writing advice when I first began writing the "Puzzling Adventures" column in *Scientific American*.

Michelle Press, Mark Alpert, and Ricki Rusting have been excellent editors at *Scientific American*, pointing out ambiguities and, often, greatly improving the exposition. Jon Erickson and Deirdre Blake have played this same role at *Dr. Dobb's Journal* for which I am very grateful. The readers of both magazines have sent in many excellent ideas.

For their continuing good energy and ideas, I'd like to thank Andy Liu and Ted Lewis. Their work with Dr. Ecco in Alberta shows that even the most eclectic puzzles can, in the right hands, convey great pedagogical value. Some of the quotes came from a compilation collected by Dr. Gabriel Robins (the reader will eventually understand that this is a hint). My friend and writer Loren Singer continues to serve as a source of inspiration.

Others, including David Molnar, Carl Bosley, Tom Rokicki, and Ariana Green, have helped with cryptography or writing. Arthur Whitney's beautifully conceived programming language K has never let me down.

My editors Bob Weil and Brendan Curry, copyeditor Carol Rose, production manager Anna Oler, and designer Mary McDonnell have been terrific. Norton is a great publishing house thanks to such uniformly talented and hard-working people. The whimsical drawings come from the unfettered hand of Gary Zamchick.

Finally, Tyler, Cloe, and Karen have struggled with and finally untied many of these puzzles.

Logic

SHIFTY WITNESSES

"XBA'G NBARELGHhLGP iP QBE MPNBiTARGSP NSLTEiLA.
c'i fHFG GSP NSTPQ QTEP SjOELAG."YEAPFGB
    XE. YNNB iLj SLIP FHRRPFGTBAF QBE L JLhgTAR
GBHE.WBiP CEPCLEPO.

Skipping the preliminaries, the detective stated his problem: "We have five witnesses whom we don't trust. The five have trailed a group of 10 suspected drug dealers. For each suspect, they take a vote about whether the suspect has drugs. Here is a summary of the votes:

Suspect 1: all five vote 'has drugs.'

Suspect 2: all five vote 'has no drugs.'

Suspect 3: three vote 'has no drugs' and two vote 'has drugs.'

Suspect 4: all five vote 'has drugs.'

Suspect 5: four vote 'has drugs' and one votes 'has no drugs.'

Suspect 6: all five vote 'has no drugs.'

Suspect 7: three vote 'has drugs' and two vote 'has no drugs.'

Suspect 8: all five vote 'has drugs.'

Suspect 9: all five vote 'has no drugs.'

Suspect 10: four vote 'has no drugs' and one votes 'has drugs.'

"Can you tell us which suspects have drugs given only that the total number of lies among all witnesses is eight or nine and most of the lies claim 'has no drugs' when the truth is 'has drugs'? They are a corrupt bunch."

WARM-UP.

What is the smallest number of lies there could be judging only from the accusations?

SOLUTION to WARM-UP.

Every non-unanimous vote (disagreement) must correspond to a number of lies at least equal to the minority view and perhaps the majority view. As we can see there are four occasions of disagreement: three against two occurs twice and four against one occurs twice. Adding up the minority views gives us six lies.

TERMINATING LEAKS

c VXGXUkbmr bl JIX paJ bl FIJpI Ur HTIr KXJKGX aX bl
ZGTW aX WJXII'm FIJp.A.3. 4XIVFXI (tll0-t2xy)
MaX TnmaJk pbGG HXXm rJn bl maX HJkIbIZ.

The celebrity governor of a certain sun-rich state shares confidences with nine advisors. To his dismay, however, some of his most intimate (and, sometimes, crude) thoughts have lately been appearing in the newspapers the day after he reveals them. A common technique for discovering leakers is to tell each suspect some unique piece of information (a tidbit) and then see if it spreads. But the governor discovers that this approach will not be good enough: newspaper editors will print a story if and only if at least three advisors attest to the tidbit. He is quite sure there are no more than three leakers. He has a dilemma. If he tells a tidbit to everyone, it will certainly be reported, but he will not have learned anything. If he tells a tidbit to one or two people, it won't be reported. He can choose a different tidbit for each triplet of people, but nine confidants can form 84 triplets. That's just too many. The governor wants to discover the leakers faster.

He arrives at the following strategy: he will tell tidbits to foursomes, a different tidbit to a different foursome each day. Once a leak occurs, he will ask questions of triplets within the guilty foursome to discover the guilty triplet. One of his goals is to provoke no more than two additional leaks—one from a foursome and, at most, one from a threesome. Another goal is to discover the triplet using at most 25 tidbits.

WARM-UP.

Suppose the governor tells a tidbit to advisors 1, 2, 3, and 4 the first day without a leak, and a second tidbit to advisors 2,3, 4, and 5 the next day with the same outcome. But his secret third tidbit, told to advisors 1, 2, 4, and 5, leaks. Which triplets are suspect? Think before you read on.

SOLUTION to WARM-UP.

Only two of the four that could be formed from the third quartet (1, 2, 4, 5) are suspect: 1, 2, 5 and 1, 4, 5. If either of the other two triplets (1, 2, 4, or 2, 4, 5) comprised the leakers, it would have leaked its tidbit during the first two days. Because the governor knows there can be only three leakers, he needs to test only one of the remaining two suspect triplets.

1. Can you help the governor guarantee to find the leakers using 25 tidbits and two leaks in all, assuming he follows the above strategy?

The governor might be able to find the precise triplet using far fewer tidbits, if sometimes he were to spread tidbits to more than four people and if he were willing to tolerate more than two leaks.

2. Can you find the leaking triplet using fewer tidbits under these conditions, perhaps tolerating more leaks? (Hint: This requires fewer than 10 tidbits.)

3. What if the governor were willing to ask more than four people, but still wants only two leaks? (Hint: You should be able to get under 15.)

COMPETITIVE CLAIRVOYANCE

"Eb ukq YWj Ykqjp ukqn Kkjau, ukq Zkj'p dWra W XeJJekj
ZkJJWno." - F. LWqJ Cappu (w45x-w532)
Pda KeJJajeqK Xacejo sepd W psk.

Suppose you have just taken up squash and plan to play several times a week. You know you are prone to injury. There are two payment plans: a yearly membership that costs $400 and entitles you to unlimited use, or a $20 pay-per-use. How many times should you pay per use to guarantee that you won't regret your expenditure of money too much?

To make this precise, consider the situation this way: you will play every day until you get injured and then you won't play anymore. A clairvoyant oracle would know when and if you would get injured. The clairvoyant oracle would either purchase use passes or yearly memberships. You want to minimize the amount you spend divided by the amount the clairvoyant oracle would spend. Call that the regret ratio.

If you decide to buy a yearly membership right from the start and then get injured on your first day on the court, your regret ratio would be 20—the $400 you spent divided by the $20 that the oracle would have spent. If you decide to pay for each use for a whole year and play 100 times before getting injured, your regret ratio would be 5—the $2,000 you spent divided by the $400 that the oracle would have spent.

WARM-UP.

Is there a way to keep your regret ratio below 2 no matter when you get injured? Think before you read on.

SOLUTION to WARM-UP.

After you have paid 19 times, buy a membership. Your regret multiple will be less than 2. Here's why. If you don't injure yourself the rest of the year, you would pay less than twice as much as the clairvoyant oracle would have done by buying the membership immediately. If you injure yourself in the game after buying the year membership, you've paid less than twice as much as if you had just bought use passes the whole time.

This form of "competitive analysis" as it's known in the trade can be applied to many fields. Should you pursue a career in areas X and Y that are both safe or in X and Y' where Y' is quite risky but is the field you are passionate about? (If you value your passion high enough but are not independently wealthy, then X and Y' have been known to work.)

Let's try a slightly more quantitative example. You have 90 tickets that you can trade for payoff cards which you will be offered over the next two hours. There is a ticket exchange booth in which a man has a pile of $1 and $5 bills. When you approach the ticket exchange booth, the man inside will offer the top bill—either $1 or $5—for each of your tickets. You have the option of rejecting a $1 offer in the hope that the man will later proffer $5 for the ticket. But if you reject a bill, the man will put it aside and you will never see it again. Moreover the man has the right to halt the trading at any time. After that point the tickets are worth nothing. A clairvoyant oracle would know in advance what the man will offer and when he will stop, but you don't.

1. Can you find a strategy that will guarantee a regret ratio (in this case, the clairvoyant's winnings divided by your winnings) that is no more than 1.8?

2. What would be your strategy if the two possible offers were $1 and $1 million? Does the regret ratio improve or worsen?

3. Can you find a general solution for any pair of offer values?

4. What if there were three or more offer values, say u, v, w in increasing order?

GOING SOUTH

Gr gq qYgb rfYr rfmqcufm lcejcar fgqrmpw Ypc bcqrglcb rm pcncYr gr, Zsr G bml'rslbcpqrYlb ufw rfcw bml'r Jsqr bpmn rfc ajYqq Ydrcp rfc dgpqr osgx.

Qm ufYr bm uc bm ugrf rfgq YlgkYj, rfgq afgknYlxccrfYr fYq amloscpcb rfc njYlcr Ylb qcckq Zclr ml psglcgle gr?8cr'q qrYpr ugrf ufYr uc ksqr bm Ylb rfcl umpK ZYaKuYpbq.

Gr gq Y uccKclb bYw.

You are planning a solo expedition to the Dry Valleys in Antarctica. You want to take flares with you in case you get in trouble. The 100-mile-per-hour Katabatic winds require special equipment of all sorts. Unfortunately, the flares may have been mis-manufactured.

There are two piles of flares in front of you. Call them A and B. Each pile consists of six flares. One of the two piles has three bad ones in it. Call that the bad pile. Unfortunately, you don't know which pile is bad at first. In fact, all the flares look equally good.

To test a flare, you must light it. Lighting it will tell you whether it is good or bad, but even if good, you won't be able to use it again.

WARM-UP.

How could you be sure to get two good flares and waste at most four, assuming the bad pile has three bad flares? Think before you read on.

SOLUTION to WARM-UP.

Start testing from one pile. If you find a bad, then stop and take your two from the other pile. If you take four and they are all good, then the remaining two in that pile are good.

That was just a warm-up. In the actual problem, after your testing, you want to collect five flares and take them to Antarctica, where you want all of them to work. You don't want to carry any extras.

1. Can you find a testing method that will give you a probability of 3/4 or better that you can take five flares that will all work and that requires testing as few flares as possible?

2. What would be your testing strategy if you needed five good flares but there were four bad flares in the bad pile? What would your probability of success be?

3. Returning to the case where there are three bad flares in the bad pile, what would your strategy be if you needed seven good flares? What would your probability of success be?

4. If there were four bad flares in the bad pile but you needed seven good flares, then what would your strategy be?

5. Is there anything strange going on here?

6. In the case where the bad pile has three bad flares, you could still take all 12 and get five good ones. You could even take five from each pile and be sure to get five good ones. Can you design a method that will enable you to take no more than seven and have five good flares no matter how the flares are distributed? How about a strategy to enable you to take no more than eight?

7. Suppose that you insist on guaranteeing that you never have to carry more than the number you found in your answer to the previous question and further guaranteeing that you have five good flares. How can you nevertheless guarantee that most of the time you need only carry five with you?

THE COLOR FAIRIES

"uXe19a5 2b4C 31e5: R5g e5fg, 3bDC ebbZf, g12Y5
2b4CAbeX, f3eh2, fge5ff e5Y956 Dba5." L4i5eg9f5Z5ag
t85 Zbag8 41g5 9f b44.

In a mythical land, fairies visit children at night and leave them pearls. But each fairy is attracted only to a particular color. Suppose, for example, that a fairy named Liz is attracted to the color aqua. She will visit each child and will leave a pearl on that child's bedside table for each aqua star she sees above the child's head. If another fairy named Maria who is attracted to crimson flies with Liz and sees a child with a crimson star, she will leave a pearl on the child's table. So, a child having both crimson and aqua stars would find two pearls.

Here is what you know:

- The fairies' names are Cloe, Ariana, Oliviana, Anya, and Caroline.
- The colors are silver, sage, gold, rose, turquoise, ivory, violet, emerald, and earth.
- At least one fairy likes turquoise and one likes earth.
- The children are Tyler, Jordan, and David.
- Tyler has rose, turquoise, and violet stars above his head. Jordan has sage, violet, and ivory. David has sage, violet, and emerald.

Here is what happens:

- The first night Anya, Caroline, and Cloe fly in and leave one pearl for Tyler and Jordan each and two for David.
- The second night Anya, Caroline, and Oliviana fly in and leave no pearls for Tyler, but two for Jordan and David each.

- The third night Anya, Caroline, and Ariana fly in and leave no pearls for Tyler, but one for Jordan and two for David.
- The fourth night Cloe, Ariana, and Oliviana fly in and leave one pearl for Tyler and Jordan each and none for David.
- The fifth and last night Anya, Cloe, and Oliviana fly in and leave one pearl for Tyler, two for Jordan, and one for David.

Which fairy is attracted to which color?

ALTERNATING LIARS

Ri frxuvh, wBh whpswdwCrq Cv wr eodph vrphrqh,dqrwBhu
uhoCdeoh Bxpdq hprwCrq.Vrph zdqw wr uroo edfn wBh forfn.Qr pruh
vrfCdo vdihw2 qhw.VxuyCydo ri wBh iCwwhvw.Xqohvv p2 frpsdq2
qhhgv d vxevCg2 wBdw Cv.IdCuqhvv vwrsv zCwB ph, p2 idpCo2, p2
Aurxs.
    jw oChv zCwBCq wzr eorfnv ri ZdvBCqAwrq Vtxduh Sdun.

You have caught five members of a very evil gang. You have only a
few minutes with them. They have been trained to be alternating
liars (you'll see what this means). One of them, however, is your side's
agent, who will always tell the truth. You want to find out which
one—fast.

You find the five in a room. They fall into two types: a consistent truth teller, who will answer every question truthfully, and alternating liars, who alternate in the truthfulness of their answers: . . . ,
true, false, true, false, true, false, ... Unfortunately, you do not
know whether any of the alternating liars will start by telling the
truth. In fact an alternating liar may decide whether to answer your
first question truthfully or falsely after he hears the question. But after
his first answer, he must alternate truth with lies. You also know that
everyone in the room knows who the consistent truth teller is.

Now, you are to determine who the consistent truth teller is. All
the suspects look equally trustworthless (if you will excuse the
coinage). However, you are allowed only two questions. You must
address each question to one person (perhaps a different person in
the two questions) and you will be answered by that person only.
(You are not in the room when you ask these questions, as they are a
scary lot.)

WARM-UP.

Find a way to determine who is the consistent truth teller if there were only three people in the room and you had three yes/no questions to ask.

SOLUTION to WARM-UP.

Label the people A, B, and C. Ask A: "Are you a consistent truth teller?" Then repeat the question to A. If A answers yes both times, then A is the truth teller. If A answers yes the first time and no the second, then A is an alternating liar and his next response will be a lie; so, you would ask A whether B is a consistent truth teller and would believe the opposite of whatever he says. If A responds no the first time and yes the second, then A is an alternating liar and his next response will be the truth; so, you would ask A whether B is a consistent truth teller and would believe him.

1. *You have five people to choose among, you are allowed to ask only two questions (of the same or different respondents). One must be a yes/no question, but the other may ask the respondent to point to at most one person. Can you find the consistent truth teller?*

2. *Suppose you were allowed to broadcast your questions to all five people and get an answer from each one. But the questions have to be yes/no. Could you do it with two broadcast questions?*

3. *How many questions would you need if there were seven people in the room?*

4. *Suppose there were only two people in the room, one a consistent truth teller and one an alternating liar. Could you determine who was who with one yes/no question to one respondent?*

ALPHA MALOS

"n3 KPK j47 G3 KPK 2GzK8 91K N141K N471J
H1m3J."bGIG92G tG3JIm

hKI43J, 4L7 7K1G9m4381m5 94 43K G3491K7:NK IG3 34
143kK7 GI9 1mzK NG77m3k I1G38.b489 7KM41L9m438 IGMK
7K51GIKJ I4143mG1m82 HP 1477mH1K JmI9G94781m58.CG78 4j
1mHK7G9m43: Jm994.CK IGMK G k7KG9 455479L3m9P 94
K8IG5K 91K IPI1K 4j Mm41K3IK.d3K 1K8843 4j lm8947P m8
91G9 8N473 K3K2mK8 IG3 HKI42K j7mK3J8Gj9K7 91K 5G88m3k
4j G kK3K7G9m43.ilK p41J CG7 3KMK7 HKIG2K u49 G3J 34N
m9'8 I1KG7 91G9 m9 3KMK73KKJKJ 94.rMK3 K3K2mK8
7KG1mQK 91G9 91KP IGMK 2LII m3I42243−14MK 4j Ilm1J7K3,
JGP8 G9 91K HKGII, G3J HKGL9mjL18L38K98.i148K 8L38K98
2L89 349 k4 GNGP.

v9 1mK8 3KG7 G I473K7 Nm91m3 j4L7 H14Iz8 j742
bGIJ4LkG1 h9.

In a certain jungle culture called the Machonamo, men fight over women. Sometimes the fights become violent. If a man A suspects that some other man B has had sport with A's woman (I'm trying to give as accurate a translation of Machonamon as possible, including the use of the possessive), A will kick B.

Suspicion comes from accusations by third parties. C may tell A that B sported with A's woman. At that point A will kick B. He is "honor-bound" to do so. When kicking B, A will tell B it was C who accused B. A must be honest about this. When B hears this, if the accusation is false, B is duty-bound to kick C. If the accusation is true, B will not kick C for this reason (by tradition and for fear that C may come up with proof that would cause B's banishment). Note that several men might accuse someone of sporting.

Since these kicks lead to some bad feelings, most women in fact remain monogamous. It's just that people get to talking . . .

An anthropologist interested in marital taboos visited the Machonamos recently, but did not know the language. Our job is to interpret her journal.

WARM-UP.
Suppose the following happened:

B speaks to A,

C speaks to A,

A kicks C,

C kicks B,

A kicks B.

SOLUTION to WARM-UP.
Here the only interpretation is that B sported with A's woman. That explains why C kicked B and why B never kicked C.

Now here is the record in the anthropologist's journal:

C speaks to F,

B speaks to E,

A speaks to F,

F speaks to A,

E speaks to B,

A speaks to B,

F kicks D,

E kicks A,

A kicks C,

A kicks B,

B kicks A,

A kicks E,

B kicks C,

C kicks A.

Based on these journal entries, determine who sported with whom.

DISPOSABLE COURIERS

"y oJPI j751 N7 Nol K14plm NoJN Nol Qo741 Q7L4k pM J6
16pn5J, J oJL541MM 16pn5J NoJN pM 5Jkl NILLpK41 KS 70L 7Q6
5Jk JNN158N N7 p6NIL8LIN pN JM No70no pN oJk J6 06klL4Sp6n
NL0No." - D5KILN7 uj7
    BII S70 p6 Nol M8Lp6n. BII S70 p6 Nol Pp44Jnl.

Any message can be encoded into ones and zeros, so any secret can be represented as a number. For example, the word *meet* in "Be sure to meet your contact at the corner of Constitution and Lake" would be represented in most computers by 109 101 101 116 in decimal, using the encoding known as ASCII.

Suppose you want to send a secret message, say a rendezvous time and location, via five couriers, but you fear that one or two will be caught. You therefore want to spread the secret among the couriers such that any three of them can together reconstruct the secret but two or fewer cannot. Because messages are encoded numbers, you could think of this problem as having the five couriers share the secret of a number. Intuitively, you might think it wise to give each of them part of the number, but that's not the most secure approach. Instead, you want to establish an information cliff: two couriers give you no useful information at all, but three give you the entire message. To reach that goal, you have to think of a more clever plan.

WARM-UP.

Suppose I play a game in which I think of a point in a plane—say the point (13,6)—and ask my friends to identify that spot. As a clue, I give Martha the line $x = 13$, Jaime the line $y = 6$, and Valeria the line $y = x - 7$. Now, before they talk to one another, no friend has any useful information.

SOLUTION to WARM-UP.

As far as each one knows, the point could be any one of the infinite points in his or her line. On the other hand, any two of my friends can find the answer instantly by determining where their lines intersect. The intersection uniquely defines the point. My friends go from facing infinite possibilities down to achieving precise knowledge (see Figure 1).

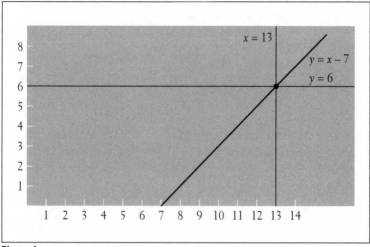

Figure 1

Similar reasoning will give you a solution to the five-courier problem.

1. There is a kindred warm-up problem in which the secret is a line. Which information would you then give each of your three friends?

2. Now try either approach for the five-courier problem.

THE DELPHI FLIP

"qj38178: F 4J6732 136J 128J6J78JI 12 k117JzK 8kF2 12
1J." - m1G637J nIJ6HJ (QXTR-QYQT)
    hkJ F98k36 MIzz MJF6 IF6y 792jzF77J7.

Predicting the future accurately is most useful in betting games—the
stock market comes to mind. Unfortunately, perfect oracles are hard
to come by (the stock market comes to mind, again). This puzzle
considers how to take advantage of the flaky oracles (dare I say stock
brokers?) one is likely to find.

You have $100 to start with and 10 bets to make. Each bet turns
on the result of a coin flip. The oracle will tell you which way the
coin will fall, but may lie on just one occasion and may do so after
seeing your bet for that flip. You can find a counterparty who will
give you even odds on any bet you make, so placing an x dollar bet
will return $2x$ dollars to you if the oracle tells the truth about that flip
and, well, zilch if the oracle lies.

1. *How do you end up with the greatest possible final amount, no matter when the
oracle chooses to lie? How would that change if you had 20 bets?*

2. *Suppose you have to decide the amount of all of your bets in advance without
knowing when the oracle will lie. What should your bets be in that case and what
final amount can you be sure to get no matter when (and if) the oracle chooses to
lie? (You lose everything if you plan for a bet on a particular move but end up having too little money at that time.)*

WARM-UP.

Suppose there are three flips and at most one lie. You have $100.
How much should you bet the first time? Given the outcome, how
much should you bet the second and third times? Figure 2 shows
some good alternatives.

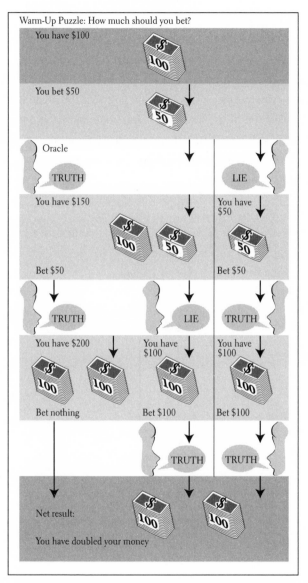

Figure 2 Decision graph showing how to guarantee to go from $100 to $200 after three bets given that the oracle will lie at most once.

SOLUTION to WARM-UP.

You bet $50 on the first flip. If the oracle tells you the correct answer on the first flip, then you bet $50 again and if the oracle lies to you on that flip you bet all $100 you have left on the third flip. If the oracle tells you the correct answer on the first two flips, you bet nothing on the last flip. If the oracle tells you the incorrect answer on the first flip, you still have $50 left and you bet all your money the second and third times. No matter what happens you'll have $200 at the end.

HINT.

Clearly, until the oracle has given you his malicious advice, you should never bet everything on one flip. Once he has, you can keep doubling your bet.

CROWNING THE MINOTAUR

"Eqn OPNrlP8n00 9o Pqn 6jS0 r0 7jmnknjNjk6n kU Pqn ojlP
PqjP 89 98n jkrmn0 kU Pqn7"Dj6PU59R-DqlqnmNr8, j 6jPn WePq
ln8PQNU CQ00rj8 SNrPnN.

xrNOP, 9QN Nn6jPr980qrL P9 8jPQNn:Sn LNnOnNRn Pqn
L6j85P98 r8 Pqn 91nj80 j8m Pqno9NnOP0 98 6j8m 09 Sn lj8
198Pr8Qn P9 kNnjPqn.Hn n80QNn PqjP Pqn SjPnN r0 16nj8
n89Qpqo9N QO j8m o9N Pqn or0q Sn njP.Hn LNnOnNRn n89Qpq
kr9mrRnNOrPU o9NPqn OjonPU 9o 9QN IN9LOj8m o9N Pqn Oj5n 9o
Pqn p99m rP m9n0 P9 9QN 09Q60. Hn LjU o9N N9jm0 PqN9Qpq
PjTn0, SqU 89P 8jPQNn?wjNPq-Srmn PjTn0 0q9Q6m oj66 r8
LN9L9NPr98 P9n8RrN987n8Pj6 mj7jpn ljQ0nm kU
7j8QojlPQNr8p,nTln00rRn Lj15jpr8p, 9N nTln00rRn
PNj80L9NP.Eqn 798nU 0q9Q6m kn Q0nm o9N 16nj8QL, o9N
S97n8'0 NrpqP0,j8m o9N oj7r6U L6j88r8p.

Eqn lrPU r0 hnS J9N5 urPU."

Early in the life of the ferocious Minotaur (the always hungry half-man half-bull imprisoned in Daedalus's labyrinth), King Minos of Crete spoke to three of his youthful prisoners. "You know that you will die if you fight the Minotaur unarmed. I propose therefore a chance at reprieve. I will separate and blindfold you. Then, on each of you, I will place either a red crown or a blue crown. I will choose the color for each person by flipping one of my lovely Cretian coins, which you can assume to be fair. I will then place you at the statues of Athena, Poseidon, and Zeus which are evenly spaced around my lovely stadium. You will be surrounded by a screen that will permit your fellows to see your crown and you to see theirs once the guards remove your blindfolds. The screen will prevent you from sending and receiving any signals, so you will not be able to communicate with one another once you are in the stadium. (A guard standing next to you will cut off your head if you try.)

"Here is the proposition: After your guard removes your blind-fold, each of you will have 10 seconds to tell your guard either 'blue', 'red', or 'pass' concerning the color of your crown. Fifteen seconds after removing your blindfold, your guard will give a thumbs up if you guess correctly, a thumbs down if you guess incorrectly, and a flat hand if you say pass. (Note that all guards will answer simultaneously, so you prisoners cannot communicate by timing.)

"If you all say pass, you must enter the labyrinth and fight the Minotaur. If all who don't pass are correct, then you all go free. If anyone is incorrect, then again you all face the Minotaur. If any of you tries to signal another, those still living get their heads chopped off."

ШАRM-UP.

What is the probability that the prisoners will win if they all bet?

SOLUTION to WARM-UP.

Only 1 in 8, since each has a probability of 1/2 of being wrong each time.

Now, you may think that you have only a 50 percent chance of surviving, by simply designating one prisoner as the guesser. But if you are clever you will realize that you can design a strategy such that the prisoners have a 75 percent chance of winning. The strategy involves a rule that each prisoner must follow, but that requires no communication among the prisoners once they are in the stadium.

1. Would you like to give it a try?

If we change the rules so that each prisoner can "bet" zero or more points about the color of his crown, the prisoner team wins or loses that many points depending on whether he is right, and the prisoner team wins if the team wins more points than it loses, then the chance that the prisoners win may improve.

2. Can you design a strategy to make it so?

3. *Using the original rules as stated by the King, but with seven prisoners, can you increase the odds beyond 3/4? (You may assume that each prisoner is going to stand underneath a statue from this collection: Apollo, Aries, Athena, Hera, Minos, Poseidon, and Zeus.)*

4. *What if you have seven prisoners and they can play for points as in question 2.*

SAFETY IN DIFFICULTY

"Fc F tbob qtl-cXZba, tlrKa F Yb tbXofkd qefp lkb?" -
gYoXeXj 7fkZlKk (x506-x532)
    Qeb bsbkq qXJbp mKXZb fk jXv.

Number theory, once thought an esoteric discipline concerned with curious properties of prime numbers, turns out to form the underpinnings of modern cryptography. The Rivest-Shamir-Adleman public-key cryptography algorithm (RSA for short) used in most e-commerce transactions is based on the (unproved, but widely believed) difficulty of factoring a number that is the product of two primes.

Multiplying two large primes is an example of a so-called one-way function, because the multiplication takes only a few microseconds and the time is proportional to the length of the binary representations of the numbers. By contrast, discovering the two primes given the product is slow, taking hours for a 512-bit product. For 2048-bit numbers, factoring is considered impractical, as far as is publicly known. Fast factoring algorithms, if they exist, would have untold applications for industrial and even military espionage.

This brings us to a puzzle first posed by John McCarthy (inventor of the programming language Lisp and theoretical Artificial Intelligence) and solved by Michael Rabin (the playful inventor of so many important computer algorithms) in the 1950s. The puzzle goes like this: You have a bunch of spies ready to go into enemy territory. When they return to cross the frontier into your country, you want them to avoid being shot, while at the same time preventing enemy spies from entering. So each must present a password to the guards which the guards will verify. Whereas you trust your spies and your guards are loyal, you believe the guards may loosen their tongues in bars at night.

What information should the guards receive and how should the spies present their passwords so that only your spies get through—even if the guards go out for a couple of pints? (Consider the discussion about primes to be a hint.)

Graphs and Circuits

FISHY BUSINESS

DBtfu qh c hgcvAgt cnyclu gpf wr dgBpi rBigqpu...
Kv nBgu yBvABp vyq dnqemu htqo UwnnBxcp uvtggv.

The resort was far too charming a place for such a senseless, smelly
crime. Seven rustic bungalows, some of them on a lagoon (A through
D) and some on the oceanfront (F and G), were connected by pathways as shown in Figure 3. A fisherman saw a sinister-looking man
approach the resort from the lagoon carrying a large basket and
sneak into one of the bungalows bordering that body of water. This
man then stalked along the pathways from one bungalow to the next,
leaving rotting fish everywhere.

Police detectives determined from a set of muddy footprints that

Figure 3

the madman had traveled along each pathway exactly once. The detectives saw no footprints leading away from the resort, so they concluded that the fish vandal was still hiding in one of the bungalows. Unfortunately, the footprints on the pathways were so indistinct that the detectives couldn't tell the direction in which they pointed. What is more, the fisherman couldn't remember which of the four bungalows on the lagoon the vandal had first entered. The police were therefore unable to retrace the deviant's route—but they knew was that he'd never walked the same pathway twice.

1. *Can you find the one bungalow in which the fish vandal must be hiding?*

2. *What if the fish vandal could retrace at most one path but must still cover all paths and still come in on the lagoon side. Then where might he be?*

PERFECT BILLIARDS

"bFF IjmIZ JSkkWk IZjImYZ IZjWW kISYWk. gajkl, al ak
jaVaUmFWV. 9WUIHV, al ak naIFWHIFq IJJIkWV. LZajV, al ak
SUUWJIWV Sk TWaHY kWFX-WnaVWHI." - bjIZmj
9UZIJWHZSmWj (syzz-szx0)
    Al FaWk oalZaH IHW TFIUE IX cFWWUEWj 91.

Imagine that you are playing pocket billiards on a pool table that is
three meters long and one meter wide. The table has been engi-
neered to perfection: when a ball banks against a cushion, the angle
of incidence is exactly equal to the angle of reflection. Furthermore,
the table is oriented so that its short sides run north-south and its
long sides east-west. The position of a ball is denoted as (x,y), where
x is the distance east from the table's southwest corner and y is the
distance north from the same corner. The table has a pocket in each
corner, but no side pockets.

WARM-UP.

Suppose you want to hit a ball at position $(2,0)$ into the southwest
pocket—that is, position $(0,0)$.

SOLUTION to WARM-UP.

The simplest strategy would be to bank once against the opposite
side cushion. Just hit the ball northwest (a slope of -1) and it will ric-
ochet off position $(1,1)$ and glide into the pocket. It's also easy to see
how to make the shot with three reflections against the cushions. Hit
the ball north-northwest at a slope of -2. The ball will bounce off
positions $(1\frac{1}{2},1)$, $(1,0)$, and $(\frac{1}{2},1)$ before dropping into the hole (see
Figure 4).

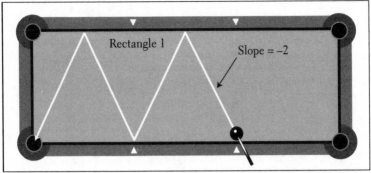

Figure 4 In going from (2,0) in rectangle 1 to (0,0) with three reflections, we are traveling two vertical for every one horizontal, so slope is -2 north by northwest.

1. *But what about making the shot by banking the ball twice against the cushions? Suppose that someone is willing to wager a sizable sum that you can't pull off this trick. At what slope should you hit the ball? (There may be several solutions.)*

2. *For those who are particularly strong in geometry, consider what would happen if the ball were initially at some arbitrary point on the southern edge. How would you discover the angle to shoot using only the ability to draw line segments and parallel line segments and to locate the midpoint of a line segment? (That is, it can all be done with a straight edge and a compass.)*

GHOST CHOREOGRAPHY

"oN98 Ni fb84i4dA. R84AM Ni f84689Bb. WA'i AM8
Ah4diNANed AM4A'i AheB5b8iec8." - Wi446 0iNceC
wM8 F84h M4i 4 9NC8 eh ceh8 Nd NA.

Her white cape undulating slowly behind her as she walked, choreographer Jane-Phillip Arlene entered Ecco's apartment. "You've heard

of ghost writers," she said. "I need a ghost choreographer." He motioned her to sit. Here I paraphrase her tale.

The dance company consists of 12 men (each represented by a figure with a hat in Figure 5a) and 20 women (each represented by a figure with a skirt). At a key point in their dance, they go from a configuration where the men are above and below the women to one where the women surround the men (see Figure 5b). In the transition, each dancer takes three steps. Each step can be either to the left, to the right, forward, or backward, or the dancer can dance in place. There are two important conditions: two dancers cannot swap positions during a step, nor can two dancers occupy the same space at the end of a step. Above all, we want to avoid collisions.

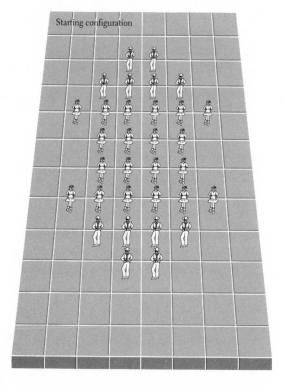

Starting configuration

Figure 5a

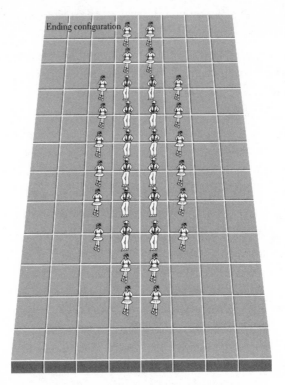

Ending configuration

Figure 5b

To see how this works, consider some slightly simpler problems. Suppose we have four x's (men) above four o's (women) and we want the o's above the x's. That is, we have

xxxx

oooo

and we want

oooo

xxxx

Imagine that they are dancing on an imaginary grid. Moving in the grid space happens at fixed time intervals. All the dancers move in synchrony. However, no two dancers are allowed to cross one another in opposite directions (directly swap places). So,

X

O

cannot become

O

X

by swapping places.

Various solutions are possible. Here is a four-step solution:

XXXX

OOOO

OXXX

OOOX

OOXX

OOXX

OOOX

OXXX

OOOO

XXXX

However, a two-step solution is possible.

1. Can you find it and make it work for any even number of dancers?

Here is the next step in complexity. Suppose we start with

xxxxxxxxx

xxxxxxxxx

000000000

000000000

and want to get

000000000

000000000

xxxxxxxxx

xxxxxxxxx

2. *How would you do that?*

The starting and ending configurations of the dancers that Jane-Phillip wants are shown below. Can you design the moves of the dancers at each step so that they can reach the final position without violating the conditions?

3. *Can you design the moves of the dancers at each step to satisfy these conditions?*

Here are the configurations (the numbers are just there for clarity):

Starting configuration

xx   (1)

xxxx   (2)

oooooo   (3)

oooo   (4)

oooo   (5)

oooooo   (6)

xxxx   (7)

xx   (8)

Ending configuration

oo   (0)

oo   (1)

oxxo   (2)

oxxo   (3)

oxxo   (4)

oxxo   (5)

oxxo   (6)

oxxo   (7)

oo   (8)

oo   (9)

Ms. Arlene left as inconspicuously as she came.

TRUCK STOP

In a European country famous for its wine, beautiful women, and very
creative workforce, a group of truckers figured out a fine way to have
their demands for lower fuel prices met. They simply blocked roads
and dared anyone to pull them out of the way. This puzzle is dedi-
cated to them.

Consider a pentagonal road network having two lanes between
each pentagonal vertex (see Figure 6a). Suppose one vehicle can
travel from one vertex to its neighboring vertex in one minute. While
that happens, no other truck can travel in either direction using the

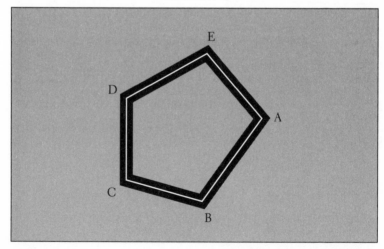

Figure 6a Original Road Network: There are two lanes between each pair of
points.

same lane. At each vertex, however, trucks can switch lanes to avoid blockage and any number of trucks can park.

In this puzzle, each vertex begins with four nonstriking delivery trucks—one going to every other vertex.

WARM-UP.

Assuming no roads are blocked, can you ensure that every truck reaches its destination in three minutes or less? Think before you read on.

SOLUTION to WARM-UP.

Outer lanes: Everyone goes right 2. This takes two time units. Then everyone goes right 1. *Inner lanes:* Similarly, but left. At the end of these three minutes, each site has sent all its trucks to all its customers. Therefore, it has received goods from all its suppliers.

WARM-UP CONTINUED.

Can you show it is impossible to do better? Try this before you read on.

SOLUTION to WARM-UP CONTINUED.

In the best case, from each vertex, one truck has to go 1 vertex away, another one 2 vertices away in the clockwise direction and similarly for the counterclockwise direction. So the total number of traversals is $5 \times (3 + 3) = 30$. There are 10 lanes in total. So, the minimum conceivable time is 3 minutes.

Now a striker decides to block the two lanes between vertices A and E (see Figure 6b).

1. In that case, how fast can the delivery trucks reach their destinations?

2. Can you prove that yours is one of the fastest possible solutions?

3. Could the strikers force the deliveries to take more time by blocking two other lanes instead?

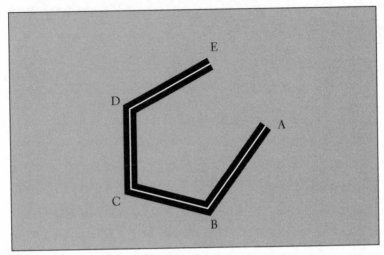

Figure 6b The striker blocks two lanes between one pair of points: A and E.

In this country, the strikers toy with the public. They slow things down rather than stopping them altogether. They call this a "social action." Sounds festive.

4. Would any two-lane blockage demand less time than six minutes?

FOLLOW-UP QUESTIONS.

What is the maximum time you could impose if you blocked three roads? How many lanes must be added to guarantee that no matter which two lanes are blocked, the whole operation takes no more than four minutes?

TRUST NO ONE

"5UR DCAI jNI ID TRI GVQ DS N IRBEINIVDC VH ID IVRAQ
ID VI." - zHPNG 8VAQR (nurq-nv00)
    e ADDi NI IUR IJOR NCQ e HRR EVPIJGRH DS KVDARCPR
NCQ QRNIU,HDBRIVBRH SVPIVDCNA NCQ HDBRIVBRH IUNI
UVTUAI HRARPIRQGRNAVII PNAARQ CRjH.e GRBVCQ BIHRAS
IUNI IURHR EVPIJGRH NGRC'I CRkI QDDG.eI QDRHC'I URAE.8R
NGR PDCQVIVDCRQ NH N HERPVRH ID IUVCi IUNI RKRGI NPI
DS KVDANRCPRjR HRR BVTUI HIGViR JH CRkI—VI'H N
PUVBENCmRR GRSARk.
    eI VH IUR SVSIU BDCIU DS IUR IRNG.

Companies buy from their competitors. Nations even buy from their
enemies. The United States Air Force flew spy planes in the cold war
that depended critically on Soviet titanium. When more than raw
material is involved, the problem of quality control becomes a challenge. In this case, the client requires a massive number of high-
voltage logic circuits to manage a switch. The supplier claims to be
reliable. Is it true?

You have received a large collection of supposedly identical circuits from a not-too-trustworthy supplier. You know which wires are
connected to which circuit elements and you are told what the elements are supposed to be. The question is, Have you been told the
truth? You want to use as few tests as possible to test each circuit.

These circuits use just two possible circuit elements: AND and
OR. Each can be characterized by a truth table relating its two inputs
to its output. AND has the following truth table:

in1	in2	out
0	0	0
0	1	0
1	0	0
1	1	1

That is, the output of the AND is 1 only when both inputs are 1.

OR's truth table looks like this:

in1	in2	out
0	0	0
0	1	1
1	0	1
1	1	1

That is, the output of the OR is 1 when either input is 1.

WARM-UP.

Suppose Figure 7 is the desired circuit. You know that the connections between the boxes are as shown, but the box values may be different (that is, an AND may become an OR or an OR may become an AND). Which test (input values) or tests could distinguish this circuit from one in which at least one box value has changed? Think before you read on.

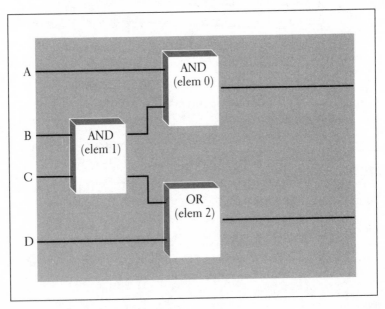

Figure 7

SOLUTION to WARM-UP.

One test is sufficient. A = 1; B = 0; C = 1; and D = 1. The top output should have value 0 and the bottom value 1. For the top output to have a value 0 when A = 1, the top circuit (elem 0) must be an AND and the input from elem 1 must be 0. Therefore elem 1 must be an AND as well. Because the output from elem 2 is a 1, it must be an OR. So, if any box changes value, at least one output will have a different value.

Now for the problems. Consider the circuit of Figure 8.

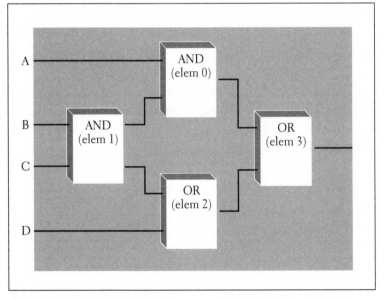

Figure 8

1. *What is the fewest number of tests you can use to verify the circuit and which test or tests would that entail?*

2. *What if the desired circuit had four ANDs?*

3. *Using the interconnection of the warm-up, which combinations of box (logic element) values can be verified by using just one test?*

PEBBLING A POLAR BEAR

"atzFr rFuty Eg2B zEBFx EFyzuxFgty, zgrBy uC zEB Eltz
yEgrr gr3g5y DruxFC5 zEB EltzBx."GCxFigt vxu2Bxh
ZEB B2Btz zgqBy vrgiB Ft zEB CFxyz ABigAB uC zEB
sFrrBtFls.

Our arctic research center is unoccupied in the winter, but its sensors continue to report on atmospheric and seismological conditions. The center consists of seven laboratory igloos interconnected by corridors. Each corridor connects two igloos. Corridors don't meet or cross one another.

We believe a polar bear has broken into one of the igloos and is wandering in the center. We will send a stun-gun team to immobilize the bear, but don't know how many gunners to send.

For safety reasons, our rules specify that at least two gunners are needed to search a room and they must approach the room together from the same corridor. Fortunately, one person is enough to guard a room that has been searched to prevent the bear from coming in. If the center were laid out in a line

o——o——o——o——o——o——o

then two gunners would be enough. They could enter at one end and go through to the other.

WARM-UP.
Suppose the center had a spoked-wheel pattern with a central igloo C and six others all connected to C and to their neighbors (see Figure 9). How many gunners would you need then?

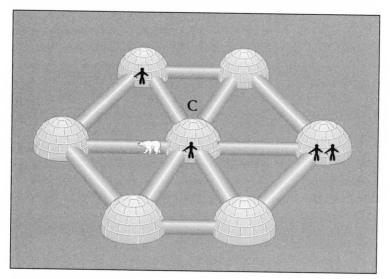

Figure 9 Spoked-wheel pattern. Here, four gunners are enough.

SOLUTION to WARM-UP.

Four gunners would be enough. Start all in the center. Send three to any outer igloo. Leave one gunner there and have the other two start moving around the circle, either clockwise or counterclockwise. Three would not be enough because the bear could always outrun the gunners and run through the center or stay on the perimeter.

Unfortunately, we don't know the precise interconnection pattern though we will when we fly in. All we can say for sure is that there are seven igloos and corridors neither meet nor cross. Moreover, there is at most one corridor connecting any two igloos. (Mathematicians would call the center's topology a planar graph.)

A polar bear would find the corridors claustrophobic, so, though it can move much faster than our gunners through them, it will not hide in a corridor. Instead it will run from igloo to igloo in essentially no time. The gunners can start at any igloo.

1. Are four gunners enough? If so, show how.

2. If not, show how many you need and that it will work.

3. Suppose there were 100 igloos arranged in a rectangular grid, where each igloo has corridors to its (at most four) grid neighbors, horizontal and vertical. What is the smallest number of gunners that is guaranteed to be enough for any rectangular grid?

THE GRAPH OF LIFE

"zn T8PM 9KMm70N 7mQmM pKI kpq51Mm7, kpK7kmN KMm
T8P R87'0 mq0pmM." uqk4 tKQm00
z0 qN jmn8Mm 0pm 0m70p 8n 0pm 6870p.

NOTE TO THE READER.
The undisguised goal of this puzzle is to shake up some scientific
biases. If you find it too technical, please come back to it at a time
when you are ready to face an open research problem.

The history of life is supposed to be based on a gigantic tree sometimes called the Tree of Life—each existing species is thought to
derive from another species that is now probably extinct. That's the
standard story, but biology eschews such absolutes. Peter and Rose-
mary Grant found that finches in the Galapagos will interbreed in
order to optimize their beak length when the weather is very dry.
Richard Lewontin showed that flies will interbreed to adjust their
adaptation to their environment. Many other studies suggest that
new species may be produced from interbreeding rather than direct
inheritance. Not always, not even usually, but sometimes. (It would
be an amusing psychological experiment to study which kinds of
maverick flies/finches/whatever would find a mate outside the
species. What would the parents think?) This puzzle explores some
mathematical implications of interbreeding on what should be called
the Graph of Life.

The genome of a species contains many genes. For example,
humans have thirty to fifty thousand—depending on whom you ask.
Many of these genes can be found with small variations on the

genomes of other species. For purposes of modelling, we divide each genome into gene groups which we name with letters. So, we have group A, B, C, and so on. Different species have variant instances of the genes in each group. For example, one species may have variant A_122 of group A whereas another species has variant A_21.

The numbers following the underbar (which we call variant subscripts or subscripts for short) have the following meaning: if X_{ij} appears in a species, then either that species or one of its ancestor species must have come from a species having variant X_i from group X. For example, if B_121 appears in (the genome of) some species S, then B_12 must appear in the genome of some ancestor species of S. We say that X_{ij} is directly derived from X_i.

Suppose we observe the following four species in nature:

S1: A_11 B_22,
S2: A_12 B_21,
S3: A_2 B_11,
S4: A_2 B_12.

Then, the following tree could explain their derivation:

```
A B
 A_1 B_2
   A_11 B_22
   A_12 B_21
 A_2 B_1
   A_2 B_11
   A_2 B_12
```

Here the indentation denotes a parent/child relationship. Thus A_1 B_2 is a parent species (the only one) of A_12 B_21 as well as A_11

B_22. This derivation and the others that follow are governed by some rules:

1. If species S has only one parent S', then every variant in S' is either directly derived from some variant in S or is the same as in S and at least one is directly derived from S.

For example, we could have

A_1 B_12 C_21
A_1 B_121 C_21

By rule 1, no tree can have the organization

A_1 ...
A_2 ...

because A_2 is not directly derived from A_1 nor can it be.

We also could not have

A_12 ...
A_1 ...

because A_1 must be an ancestor of A_12 in the tree.

2. If species S has several parents (as the result of interbreeding), then every variant of S is in exactly one parent.

By rule 2, we could have the following relationship: species 1 has A_1 B_1 C_1, species 2 has A_2 B_2 C_2, and derived species 3 has A_1 B_2 C_2.

3. If some variant of a group Z is somewhere in the tree or graph, then all the species in the graph where it is present are connected and are directly or indirectly derived from a single species having that variant.

Rule 3 basically says that a variant arises only once (at its "arise point") and after that is inherited. Consider for example:

A B
  A_1 B
   A_1 B_1
   A_1 B_2
  A_2 B
   A_2 B_1

A_1 arises once in A_1 B and after that it is derived. Similarly for A_2 and B_2. B_1 on the other hand arises twice (in A_1 B_1 and A_2 B_1). Neither instance is derived from the other. So B_1 violates rule 3.

WARM-UP.

Suppose the final species S1, S2, and S3 have the gene variants:

S1: A_1 B_1,
S2: A_1 B_2,
S3: A_2 B_1.

Can this result from a phylogenetic tree where the nonleaf nodes are extinct species derived from A B? Now try:

S1: A_1 B_1,
S2: A_1 B_2,
S3: A_2 B_3.

And now:

S1: A_12 B_21,
S2: A_11 B_22,
S3: A_22 B_12,
S4: A_21 B_11.

SOLUTION to WARM-UP.

S1: A_1 B_1,
S2: A_1 B_2,
S3: A_2 B_1,

cannot result from a phylogenetic tree as we can see by generating the candidate trees. Let's start with this one:

A B
A_1 B
A_2 B

B_1 would have to be below both A_1 B and A_2 B, violating rule 3. If we start with this:

A B
A B_1
A B_2

then A_1 would have to be below both A B_1 and A B_2, again violating rule 3.

On the other hand, here is a tree derivation for

S1: A_1 B_1,
S2: A_1 B_2,
S3: A_2 B_3.

A B
A_1 B
A_1 B_1
A_1 B_2
A_2 B_3

And here is a derivation for

S1: A_12 B_21,
S2: A_11 B_22,
S3: A_22 B_12,
S4: A_21 B_11.

A B
A_1 B_2
  A_11 B_22
  A_12 B_21
A_2 B_1
  A_22 B_12
  A_21 B_11

The warm-up suggests the following questions:

1. *Given only two groups, two subscripts per group, where each subscript can be either 1 or 2, can you add any species to the example of the last part of the warm-up and still get derivation by a tree? For example, could you add some different species of the form A_ij B_km?*

2. *What would be a smallest possible set (there could be many) of species that cannot be derived from a tree where each species is described by two groups, two subscripts per group, where each subscript can be either 1 or 2?*

3. *Find the nontree derivation of this set with the fewest interbreeding events.*

4. *The previous example had every possibility. Here is one with fewer possibilities. Are there different subsets of species which result in different trees?*

S1: A_11 B_11,
S2: A_11 B_21,
S3: A_11 B_12,
S4: A_11 B_22,
S5: A_21 B_11,
S6: A_21 B_21,
S7: A_21 B_12,
S8: A_21 B_22.

5. *Find a tree and two interbreeding events to derive:*

S1: $A\_11 \ B\_11$,

S2: $A\_11 \ B\_21$,

S3: $A\_11 \ B\_12$,

S4: $A\_11 \ B\_22$,

S5: $A\_21 \ B\_11$,

S6: $A\_21 \ B\_12$.

6. *How many species could you describe with a tree having* k *groups, two subscripts, and two values per subscript?*

7. *How about* k *groups,* m *subscripts, and* n *values per subscript?*

Research is now going on to find a general theory.

GRIDSPEED

"7WPK KWT PjKFDFQXCT WPJ GIPRKXRPCCn ITPRWTS
KWT CXDXK FU XKJ STkTCFGDTEK XJ JjVVTJKTS Qn KWT UPRK
KWPK SjIXEV KWT GPJK nTPI EF XDGIFkTDTEKJ FU P
IPSXRPC EPKjIT WPkT QTTE XEKIFSjRTS."- 6RXTEKXUXR
YDTIXRPE, hPE. q, pxOx
   bI. cRRF DPn IIXKT JFDTKWXEV SFIE KWPK nFj DPn
RWFFJT KF SF.

You are given a 6-by-6 grid of streets where the north-south streets are elevated with respect to the west-east streets. Entrance and exit ramps connect the two at intersection points. There are no traffic lights, so switching from a north-south street to an east-west street (and vice versa) takes essentially zero time. There is very little traffic but the police patrol very carefully for speeders.

From the top to the bottom the speed limits increase as follows: 10 miles per hour, 20 miles per hour, . . . , 60 miles per hour. Similarly, going from left to right 10, 20, 30, 40, 50, 60 (see Figure 10). Each grid cell is 10 miles from its nearest neighbor.

WARM-UP.
What is the fastest way to go from (1,1) to (6,3)?

SOLUTION to WARM-UP.
An optimally fast route from (1,1) to (6,3) is right to (2,1) in the first hour, then down to (2,3) in the second hour, and finally to (6,3) in the next $\frac{4}{3}$ hours. Total time: 3 hours, 20 minutes. There are others that are just as fast.

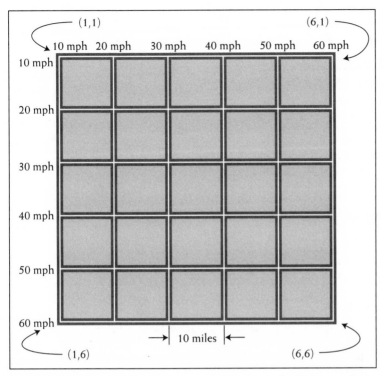

Figure 10 The upper-left-hand corner is $(1,1)$; upper-right-hand corner is $(6,1)$. The top row and left-most column allow speeds of 10 miles per hour and each grid point is 10 miles apart from its vertical and horizontal neighbors.

There are many slower ways that are still direct. For example, go from $(1,1)$ to $(6,1)$ in 5 hours and then 20 minutes to get down to $(6,3)$, for a total of 5 hours, 20 minutes (see Figure 11).

Here is the problem. You want to visit every intersection in as short a time as possible starting from $(1,1)$.

1. *How do you do it if you are forbidden from visiting any intersection twice?*

2. *Can you do better if you are allowed to visit intersections twice?*

3. *Is $(1,1)$ the best place to start to visit every intersection in the shortest time possible with the no two-visit restriction? (I'm not sure about the best place to start without that restriction.)*

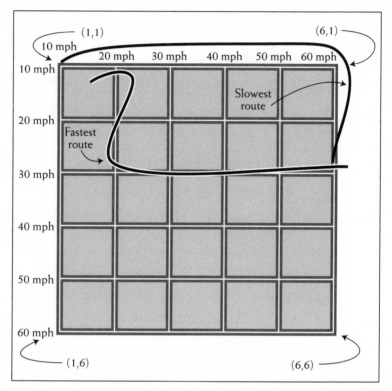

Figure 11 One of the fastest routes from (1,1) to (6,3) is right to (2,1) in the first hour, then down to (2,3) in the second hour, and finally to (6,3) in the next ⁴/₃ hours. Total time: 3 hours, 20 minutes. A slow but direct route is to go from (1,1) to (6,1) in 5 hours and then 20 minutes to get down to (6,3), for a total of 5 hours, 20 minutes.

Strategy and Games

STACKING THE DECK

Alice and Bob play a card game using the ace through seven of hearts and the ace through seven of spades. The 14 cards will be arranged somehow.

Bob thinks of a number between one and seven, then tells Alice the number. Alice deals out that many cards, finishing the first round. Bob turns over the last card dealt and Alice then deals out that many more (an ace is considered one), finishing the second round. Bob turns over the last card dealt. Again Alice deals out that many more. This continues until there are not enough cards left to deal—in which case Bob wins—or if the last card dealt in some round is the very last of the 14 cards and it is the ace of spades. In that case, Alice wins.

WARM-UP.

How can Alice arrange the cards in advance so that no matter which number between one and seven Bob chooses at the beginning, the game will end with the ace of spades as the last card turned up in a round?

SOLUTION to WARM-UP.

Here is one possible ordering with the last card being the ace of spades: 3, 4, 5, 6, 7, 6, 5, 4, 3, 2, 1, 2, 7, 1. Suppose Bob chooses four. Alice turns over the cards 3, 4, 5, then 6. After she turns over the 6, she deals 7, 6, 5, 4, 3, 2. After she turns over the 2, she deals 1, 2. She then turns over the 2 and deals 7, 1. The final 1 is the ace of spades.

Here is a distinctly harder problem.

1. *Suppose that Bob arranges all cards with face value greater than four. Then Alice, without looking at what Bob has done, arranges the remaining cards and puts her cards after Bob's on the deck. Can she still force a win?*

I've saved the hardest problems for the last. Bob takes any seven of the cards except the ace of spades. Alice may look at what Bob has done and insert cards among those of Bob's, but without changing the order of Bob's cards. Can Alice still guarantee a win? For example, if Bob arranges the cards 5, 1, 2, 6, 7, 3, 3, how can Alice win? Think before you read on.

Alice would win by inserting the cards in brackets: [7], [6], [4], [2], 5, 1, 2, 6, [5], [4], 7, 3, 3, [1]. This will guarantee her a win.

2. *Suppose Bob takes seven cards and arranges them as follows: 1, 2, 2, 3, 3, 4, 4. Can Alice still win by inserting cards among those of Bob?*

3. *But can Alice win in general?*

ANALYTICAL LINEBACKING

"ApTpP tLRpPPSNR WMSP pLp9W UspL sp tQ 9l7tLr 1
9tQR17p." - AlNM8pML vMLlNlPRp (Yedg-YfZY)

JslR oM Up lo9tPp QM 9Sns lmMSR QstLtLr 9t8tRlPW
slPoUlPp?u qtrsRpP N8lLp Rl7tLr MqqqPM9 lL ltPnPlqR nlPPtpP
tQ lL lnR Mq mplSRW lLo rPlnp.Gsp QRlRp9pLR "Gl7p RslR" slQ
nlPPtpo MTpP qPM9 1 QnsMM8WlPoQsMTp RM 1 9prlRML ltP
QRPt7p.Gsp opntQtML 9l7pPQ lPp RsMSQlLoQ Mq 9t8pQ lUlW tL
n8t9lRp-nMLRPM88poQpnSPtRW. Gsp RlPrpRQ LpTpP Qpp RsptP
7toQ lrltL. BRspPQ UsM lPp LMR (WpR) tL NMUpP oPp19 Mq
uP9lrpooML, 1 rPplRoplRs qM88MUpo mW 1 FpnMLo wM9tLr.vSR
UslR tq 1Mo qMPrpRQ Rsp lNNMtLR9pLR?

3R tQ mpqMPp Rsp RUpLRtpRs Mq Rsp 9MLRs.

Imagine a game played on a grid in which there is a very fast runner
who can make two moves in a turn, where each move is southward
or southwest or southeast.

For example, if the runner R starts at the top as shown

```
. . . R . .
. . . . . .
. . . . . .
. . . . . .
. . . . . .
. . . . . .
```

then he can end up in any of the places having a lowercase "r":

```
. . . . . .
. . r r r .
. r r r r r
. . . . . .
. . . . . .
. . . . . .
```

That is, he must move one space or two spaces, but each move is one southward and at most one east or west.

The three tackles can each move one space in any direction: north, south, east, west, northwest, southwest, northeast, or southeast. Starting here,

```
.  .  .  .  .  .
.  T  .  .  .  .
.  .  .  .  .  .
.  .  .  .  .  .
.  .  T  .  .  .
.  .  .  .  .  .
```

the northernmost one can end up in the top nine places and the southernmost one can end up in the lower nine places.

```
t  t  t  .  .  .
t  t  t  .  .  .
t  t  t  .  .  .
.  t  t  t  .  .
.  t  t  t  .  .
.  t  t  t  .  .
```

The tackles win if they prevent the runner from moving. For example,

```
.  .  .  .  .  .
.  .  .  .  .  .
.  .  .  .  .  .
.  .  .  .  R  .
.  .  .  T  T  T
.  .  .  .  .  .
```

The runner wins if he reaches the southernmost rank.

Here is the first game setup: the runner begins anywhere he wants on the northernmost row. The three tackles then place themselves anywhere that is at least three spaces away from the runner.

The runner has the first turn. After that, play alternates. Here might be an initial configuration:

```
.   R   .   .   .   .
.   .   .   .   .   .
.   .   .   .   .   .
T   .   T   .   .   .
.   T   .   .   .   .
.   .   .   .   .   .
```

1. *Can the runner or the tackles guarantee a win? If so, how?*

2. *If the tackles must start on the last row, does this change the outcome? Say how (either way).*

VENTURE BETS

Suppose you're the manager of a venture capital fund. You've identified 11 hot companies that have a decent chance of striking it rich,
providing your fund with returns 10 times as great as your original
investment in the company within three years. Your investors, however, want a safe road to riches. Can you use your mathematical
knowledge to help them make money with high likelihood?

WARM-UP.
Let's say that your fund has $17 million to invest. Each of the 11 com-
panies has a 40 percent chance of yielding tenfold returns and a 60
percent chance of going bust. These probabilities are independent
for the different companies because they are in different industries.
Your investors want at least a 60 percent probability that the fund
will grow to $60 million or more. How should you allocate the $17
million among the companies?

SOLUTION to WARM-UP.
Clearly, putting all the money in one company won't work, because
the probability of success is only 40 percent. But if you invest in two
companies, the chance that at least one of them will succeed is 64
percent (because the probability that both will fail is $0.6 \times 0.6 =$
0.36). So if you put $6 million in any two of the companies, you will
meet the required probability for a $60 million return and still have
$5 million left in reserve in case an even better investment comes
along. Spreading your bets further is not a good strategy. For exam-

ple, if you invest $3 million in each of four companies, hoping that at least two will reap a windfall, your probability of success is barely over 50 percent.

The warm-up as it stands struck some venture capitalists as too risky. By trying to achieve a 60 percent chance of success ($60 million return), it also admitted a 36 percent chance of losing all $12 million. Suppose that instead of asking for a 60 percent chance of attaining $60 million or more, you were willing to make that much with a likelihood of only about 34 percent. On the other hand, you want to reduce your chances of losing to under 5 percent.

1. How would you do that?

Now let's suppose that the economy has suddenly improved. Each of the 11 hot companies now has an 85 percent chance of yielding tenfold returns. But the demands of your investors has grown correspondingly. They want a 95 percent chance that their fund will grow from $17 million to $100 million. Again, investing all the money in one company won't work. Giving two companies $8.5 million each won't work either: although there's a 97.75 percent chance that at least one company will succeed, that yields "only" $85 million.

2. Can you find a way to achieve your investors' goals, while keeping as much money as possible in reserve?

3. What if you wanted the fund to grow to over $140 million but were willing to have a reserve of only $2 million. Then what could you do?

BLUFFHEAD

ikk sgZs lnqZkhsx hr gnfvZrg.Gdqd'r gnv cdaZsdr fds
qdrnkudc—vnqckdrrkx."Vdkk kds'r rdd.H bZm ntsrodmc xnt z0 sn z.
Xnt sghmj Zcudqshrhmf hr hllnqZk.Ne bntqrd sgd mdsvnqjr vhkk fhud
tr aZkZmbdc bnudqZfd."
    Xnt lZx ehmc lnqd hmenqlZshnm
Zsvvv.br.mxt.dct/br/eZbtksx/rgZrgZ/oZodqr/akteedqTrd sgd rZld jdx
Zr xnt ehmc gdqd.

In my gambling years, between ages 8 and 12, I used to play roulette, poker, black jack, and any other game that had a three cent ante. One of my favorite games—for its sheer purity—was what we called "Bluffhead."

To play, one person shuffles the deck. Then each person takes a card and holds it to his or her forehead. Each player can see all cards except his or her own. The highest card wins. Ace is high and suits don't matter, so ties are possible.

This puzzle has to do with inferring what players have by listening to what they say. To be concrete, suppose that Jordan, David, and Caroline each pick a card. Caroline speaks first, then David, and then Jordan. Each player says either:

"I win." (I have a higher card than anyone else.)

"I lose." (Someone else has a higher card than I have.)

"I tie as a winner." (I have the highest card, but someone else may
    have a card of equal value.)

"I don't win." (I either tie or lose.)

"I don't lose." (I either tie or win.)

"I don't know."

The players are assumed to be perfect logicians, but reveal information only through one of the phrases above. In every situation, they will make the strongest statement they know.

WARM-UP.

Suppose Caroline says, "I don't know"; then David says, "I lose"; then Jordan says, "I lose." What do we know about the cards?

SOLUTION to WARM-UP.

Caroline must have an ace and the others have less than an ace. Here's why. Caroline cannot see an ace, otherwise she could say "I don't win." Because she says she doesn't know, David realizes that he doesn't have an ace. If he sees an ace on Caroline's forehead, then he knows he loses. There is no other reason for David to know he will lose. Similarly for Jordan.

Here are some questions. See what you can infer:

1. *Caroline says, "I don't know." David says, "I don't win." Then Jordan says, "I win."*

2. *Caroline says, "I don't know." David says, "I don't win." Then Jordan says, "I tie."*

3. *Caroline says, "I don't know." David says, "I don't win." Then Jordan says, "I don't win."*

4. *Caroline says, "I don't know." David says, "I don't know." Then Jordan says, "I don't know." Then (having heard this) Caroline says, "I lose." What do David and Jordan then say and what do you know about the cards?*

Tom Rokicki has helped me work out the last one here:

5. *Caroline, David, and Jordan say, "I don't know" in order, then say, "I don't know" in a second round. In the third round, Caroline and David say, "I don't know," but Jordan says he wins. What does Jordan have and what might he see?*

ADVERSARIAL BIFURCATIONS

"gIFx DIyy3JxH CF47FFx 47y F6Jv3, q Bv7B93 vJuF 4y 429
4IF yxF q'6F xF6F2 42JFE CFGy2F." - WBF gF34 (MTUN-MUT0)
q4 vJF3 7J4IJx Gy52 CvyDu3 yG py534yx c42FF4.

In the surprising tale of espionage called "El Jardin de Senderos que
se Bifurcan" (The Garden of Bifucating Paths), Borges offers a theory
of time in which many universes exist as shadows of one another.
There is the universe in which Napoleon won at Waterloo and
French became the interlingua of earth. There is a universe in which
certain galaxies were born and displaced the Milky Way. And so on.
It's a troubling notion, neither provable nor falsifiable, neither moral
nor evil, advocated neither by science nor religion, but condemned
by neither as well. On to the puzzle.

Baskerhound, our former kidnapper, stood at the door. "You are
practical, Ecco, I know, so I wouldn't disturb you with such notions
except that my employers (whose identity I cannot reveal) have
described to me a kind of tree of bifurcating paths," he began. "Each
path has an outcome that you might interpret as the value of the
path. They are playing a kind of game against an enemy, they say.
My employers make all the moves except two and the enemies can
choose when to make each of their two moves."

Baskerhound started to sketch out a tree of possibilities.

"Let me get this straight," Ecco's 14-year-old niece, Liane, inter-
rupted. "You start at the root. You can go left or right each move.
Before making a move, you first must ask your adversary whether he
wants to choose (unless he's already chosen twice). If he doesn't, you
can choose. Please stop your drawing and let's try a warm-up."

Liane then drew a tree in indented form (you can see the equivalent in tree form in Figure 12):

A
  B
    C
      6
      177
    D
      380
      302
  E
    F
      213
      -129
    G
      618
      17

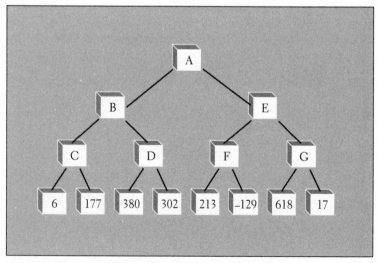

Figure 12 The game tree for Baskerhound's first problem.

"Here's how it goes," she explained. "You start at A, ask your adversary whether he wants B or E. If he doesn't choose, then you get to choose. Suppose you choose E. Then you ask your adversary whether he wants F or G. Suppose he chooses F. Then at F you ask your adversary whether he wants 213 or -129. Since your adversary wants you to get as little as possible, he chooses -129. This shows in fact that E was a bad choice for you to make. In fact, you can guarantee a win of at least 6 if you start by choosing B. If the adversary chooses E, you will do even better, since your adversary can then choose only one more move. In fact, you can then guarantee at least 17."

"You grow smarter even as you grow more beautiful," Baskerhound said. "Which is the cause of which?"

Liane appeared to ignore both the comment and the question. "Tell us the real problem, Dr. Baskerhound," she demanded.

"Maybe it's the directness," he said. "OK, here it is—a binary tree with 63 nodes. As you can see there are some bad negative numbers in this six-level tree:

A
 B
  D
   H
    P
     1055
     775
    Q
     3011
     3791
   I
    R
     -4077
     -3092
    S
     4894
     -465
  E
   J
    T
     2149
     1776
    U
     -2408
     3046
   K
    V
     18
     -1043
    W
     -3266
     2523

A
  C
    F
      L
        X
          396
          452
        Y
          2189
          483
      M
        Z
          5219
          -627
        a
          2055
          3247
    G
      N
        b
          115
          -3841
        c
          509
          187
      O
        d
          4125
          667
        e
          2338
          -6

Liane was able to show that she could win a value of 396 for Basker-hound's employers.

Can you do as well or better?

After Baskerhound left, Ecco turned to Liane. "Nice work," he said. "Do you think you would still get a positive score if your adversary had no freedom to change the numbers, but could rearrange the values of the subtree below the current node once. For example, if the current node were G, he could swap -3841 under B with 667 under D, but he would not be permitted to swap either number with -4077 under R. How well could you guarantee to do if he could perform this rearrangement only once in addition to choosing two moves?"

A pity that Ecco doesn't answer his own questions.

ULTIMATE TIC-TAC-TOE

"O3z wvxSwWVz W1 acZXZ4az 4a 1ca4V2 aXzzy e4b3
azxZzxg." - qWV 7Tvcaze4bh (iFGO-iGBi)
     O3z yZzaa xWyz 1WZ UzV 4a vVgb34V2 gzTTWe.

It was Benjamin Baskerhound once again. "Ecco, you may be nearly as surprised as I am by my new job. The FBI has engaged me to negotiate with a kidnapper. It seems they like the way I deal with adversaries."

"A replay of the theme of 'It Takes a Thief,'" Ecco said with a smile.

Baskerhound chuckled. "You wouldn't hold that small kidnapping escapade against an old friend after all these years, would you? Anyway, I have a personal vendetta against this kidnapper—they call him Sam the Snatcher. He has the conceit of being the world's greatest gamester. What nerve!

"It seems that he wants to play a game he calls Ultimate Tic-Tac-Toe against the FBI. If we win, he will charge *only* $10 million per hostage. Rates have gone up since you were my guest."

Ecco smiled again. I suspected he still missed Baskerhound's hideouts and his companions.

"So, what's the game, Dr. Baskerhound?" Liane asked.

"I'm glad someone is taking me seriously around here," Baskerhound replied and then directed his explanation to Liane. "Imagine a 4-by-4 grid (not 3-by-3). We will play a tic-tac-toe variant on this. Players alternate play as in tic-tac-toe with X moving first and then O moving second. A 'threesome for X' is a set of three cells that form a continuous line vertically, horizontally, or diagonally and that all contain X's. Similarly for a 'threesome for O'. (Note that a threesome may be contained in a foursome.) Since the grid is 4 by 4, there are

16 cells, so eight alternating turns of the two players will fill the board.

"Now for the rules: if a player gets a threesome before his last possible turn, then he loses. If he gets a threesome on his last turn, he wins. The game is of the 'sudden death' variety. As soon as there is a loser or a winner, the game stops. So if both players could get a threesome on the last turn, then the first player wins anyway. If either player loses on some turn (by getting a threesome too early), then it doesn't matter if the other player would be forced to lose on his next move. The player who loses first still loses. If neither player gets a threesome by the end of the game, the game is a draw. The big question is, Does either player have a winning strategy (that is, can force a win)? If not, can either player force a draw?

"Fortunately, we don't have to answer the big question. Sam has given us two game puzzles."

I reproduce them below.

WARM-UP.

Can X force a win from this configuration? (It's X's move.)

Figure 13

88

SOLUTION to WARM-UP.

The first move is shown in the figure below.

Figure 14 X moves in the spot X.1. No matter where O moves. X can next move in the center in such a way that X has two ways to move in its last move to get a threesome.

By case analysis nothing O can do will prevent the X-player from being able to place X in a square that will give him two opportunities to win in the last turn.

Now try this harder one:

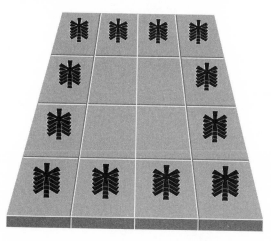

Figure 15

The asterisks represent moves by X and O. All you know is that neither has a threesome and that there are six X's and six O's among the perimeter.

Can X force a win in this case?

Still his original question weighed heavily in the air. Was a winning strategy possible for X in general or could O always force a draw? I never heard the answer.

There is another variant of the game called the Come-Back variant which is also open. In that variant, O wins if he obtains a threesome on the eighth turn regardless of whether X has already obtained a threesome on his eighth turn. On the other hand, early threesomes still face a sudden-death rule.

In that case, does O have a winning strategy?

I don't know the answer.

JUMP-SNATCH

"bxvE CA42E HAyyIwE22 6HE1E5E1 3HE8 Gx; x3HE12,
6HEwE5E1 3HE8 Gx." - X2CA1 fIuDE (LSPO-LT00)
Uxxt Fx1 2xvExwE 6I3H 1ED HAI1.

You are given a tic-tac-toe board and the following simple rules: you can jump a piece if it is between you and an empty square along any line—vertical, horizontal, or diagonal. When you jump a piece, you remove it. In the solitaire version of this game, your goal is to have only one piece left after some number of jumps.

Consider the following configuration (a dot means empty space):

.	X	X
.	X	X
.	.	.

WARM-UP.

Is there any way to ensure that only one piece will be left standing after some number of jumps?

SOLUTION to WARM-UP.

Sure. Start at the upper-right corner. Jump down. Jump diagonally up left. Jump across right. Jump diagonally down left.

Now for the questions:

1. How many spaces must you leave empty and where should they be in order to have only one piece remaining at the end?

2. Can you prove that this is the minimum?

3. If the tic-tac-toe board were 4 by 4, how many spaces must you leave empty and which one(s) to have only one remaining piece at the end?

I don't know how to generalize these to larger squares. Also, I don't know whether there exists a strategy that minimizes the number of initially empty spaces and which is "perfect" in the sense that the piece that first jumps is the only one that ever jumps and is the last one standing.

All of this leads to the hardest game I know that is played on a 3-by-3 grid and that is extremely interesting on a 4-by-4 grid. Let's call it Jump-Snatch.

Put pieces on all squares of the grid. The Snatcher removes a piece from any square he/she chooses. The Jumper then makes the first move after which moves alternate between Snatcher and Jumper until one player does a jump that leaves just one piece—thereby winning. A move consists of jumping (after doing one jump, each additional jump is optional) if a jump is possible. If no jump is possible, then the move must consist of sliding a piece toward the center. If this is not possible, then the player must slide a piece to allow a jump. The figures below illustrate the first moves of a 3-by-3 game.

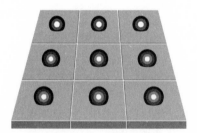
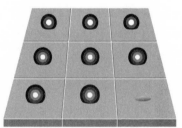

Figure 16 Jump-Snatch I: Initial setup—one piece in each grid cell.

Figure 17 Jump-Snatch II: Snatcher removes a piece. This is just an example. Any piece can be removed.

 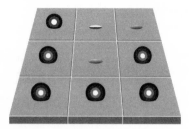

Figure 18 Jump-Snatch III: Jumper jumps over the center along the diagonal.

Figure 19 Jump-Snatch IV: Snatcher jumps over top middle.

4. *Who wins this game?*

SHORT TAPS

"fPGGn GTFGCT WPkT IFESTIUjC DTDFIXTJ QjK QPS
RFEJRXTERTJ"7WFDPJ ZIjJJXV
    bI. cRRF IXCC QT JTTE QjK EFK WTPIS.

To tap a certain microwave link, your adversary's spies must be in line of sight of the transmitter. This is dangerous so you are sure they won't do this for more than 10 minutes and at just one time.

The trouble is that you don't know which 10 minutes they will choose. You have seven messages (ranging from two to eight minutes) that you want to send in as short a time as possible. You don't want the spies to intercept four or more of them. You consider a message to be intercepted if the spies tap it from start to end; a partly tapped message won't do them any good (you send it encrypted with a one-time decryption key at the end). Each message is sent in one continuous transmission, though several messages can be sent in parallel.

Here are their lengths:

A: 2 min.,

B: 3 min.,

C: 4 min.,

D: 5 min.,

E: 6 min.,

F: 7 min.,

G: 8 min.

WARM-UP.

Suppose that you don't want the enemy spies to tap more than one complete message. Assume for starters that you start all messages on the minute. When should you send each message to minimize the total time?

SOLUTION to WARM-UP.

Here is a 36-minute solution. Send B at 0, F at 4, D at 10, G at 13, C at 20, E at 25, and A at 34. There is no 10-minute interval that contains two messages in their entirety.

Because you allow three messages to be tapped, you may be able to do much better.

1. Can you show how to do it in 15 minutes or less, again assuming that you start messages on the minute?

2. If in addition to these seven messages you had to send three four-minute messages, then can you do it in 20 minutes or less without allowing more than three messages to be tapped in their entirety and under the on-the-minute assumption?

3. To what extent can you improve your timing if you are willing to start sending messages without insisting sending be on the minute?

GRAB IT IF YOU CAN

"YC8M9: P c5E ABC89eB ENf d5ibA NOA fEe g5g9iA."W.p.
q9e7b9e (InnO-Iokl)
xN9 5i95 OA efiBN fL WfCABfe ABi99B.

In a certain form of arbitration, the plaintiff and defendant hire a retired judge and conduct a trial before that judge. The plaintiff writes down, on a piece of paper, how much he thinks he should receive, P, and puts the paper in a sealed envelope. The defendant writes down how much he should pay, D, and puts that in a separate sealed envelope. At the end of the trial, the judge makes a decision J without knowing P or D. If J is closer to the higher of D or P, then the defendant pays that higher amount. Otherwise, the defendant pays the lower amount. So, if the plaintiff thinks he deserves $18 million, the defendant thinks he should pay $0, and the judge decides the complaint is worth $8 million, then the plaintiff gets $0.

Your job is to find the strategy for the plaintiff that will give the best expected value. Suppose the judge signals that he will arrive at some number between $3 million and $10 million, but without any suggestion where. So you can assume that every dollar amount in that range has the same probability. The plaintiff is sure the judge is fair, though he is less sure about the honesty of the defendant. The plaintiff has the highest regard for the intelligence of the defendant, however.

WARM-UP.

If the plaintiff and the defendant know the judge will say $5 million and both parties know that the defendant can read what's written in the plaintiff's sealed envelope, then what should the plaintiff ask for?

SOLUTION to WARM-UP.

Also $5 million. Otherwise if the plaintiff asks for $5 million + x, then the defendant will offer $5 million − x and will pay that amount. Of course if the plaintiff asks for less than $5 million, then the defendant offers that same amount or less. So, the plaintiff should ask for and will get $5 million.

1. What should the plaintiff ask for to maximize his expected receipts in the case that he suspects the defendant can read what is in the plaintiff's sealed envelope and the judge may decide any amount between $3 million and $10 million with equal probability?

2. How will the plaintiff's request change if he knows that his sealed envelope is secure?

SKYCHASERS

"Zv n zhGB 2ulv FGt, 'IC Dy2G1D2s huB t2slGws6'. I21 uvl
Gu lFvzC 4vyBz." dvvB6 HssCu.

aFGyB, v2y yCshlGvuzFGw lv v2y D2l2yC:v2y iCzl BCzlGu6
Gz lv wyCzCy3C v2y wshuClhuB lv C5wsvyC v1FCyz.av C5wsvyC
v2y Ehsh56, lv C5wsvyC 1FC 2uG3CyzC iC6vuB.aFGz Ahu iC BvuC
Gu h 4h6 1Fhl zvhrz 2w v2y z2yws2z wyvB2A1GvuhuB sChBz 2z lv
h Avttvu F2thu huB wvzzGis6GulCy-wshuC1hy6 CulCywyGzC.

aFC hyCh Gz zv21F vD Dv2ylCCulF z1yCC1.

You have been approached by a spy agency to figure out who is sending which containers of goods to whom. Containers flow from port
to port and then get mixed up in a warehouse at each port, so you
can't trace individual containers from one end to the other. On the
other hand, you know from satellite sensing the number of containers going through each link and you know when the warehouses are
empty. Given the link flow below, and from the knowledge that
every container takes the shortest possible route to its destination,
can you determine how many containers flow from each source to
each destination?

WARM-UP.

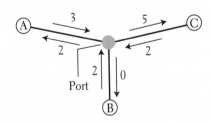

Figure 20

SOLUTION to WARM-UP.

> A to C: 3,
> B to C: 2,
> C to A: 2.

Note that B to A must be 0 because otherwise at least one container from C would have nowhere to go. Remember that containers always take their shortest path.

So much for the warm-up. Now, consider the following link traffic:

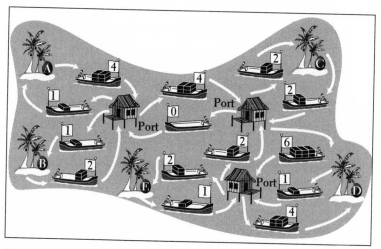

Figure 21 Contraband goods traded among countries A, B, C, D and E flow through neutral ports. Satellite cameras reveal the number of containers traveling in each direction on each link.

1. Can you determine how many containers flow from each source to each destination?

Surprisingly the answer changes quite a bit if a few shipments change by one, as shown next:

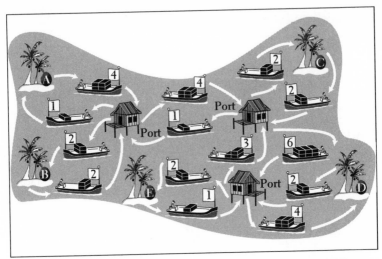

Figure 22 Contraband goods traded among countries A, B, C, D and E flow through neutral ports. Satellite cameras reveal the number of containers traveling in each direction on each link.

2. Can you say how?

NANOMUNCHERS

cH4Q Lmn41 Lmj 0582K n9 6H99n41 2n1j H K8jH3.Amn419
JmH41j 95 kH9L. x5I9 H8j H22 Lj3658H8Q.t8nj4K9 4jNj8 9LHQ
J259j IQ.A8HKnLn54 n9 3H4MkHJLM8jK IQ J533j8JnH29.Amj
HKNnJj 5k 52K 6j562j n9 8nKnJM2jK.o4 H358H2 IjHLnknJ
H4n3n93 6j83nL9 jNj8QLmn41.Amj K8jH3 mH9 IjJ53j H 35Nnj.
Amj HMLm58 0n22 3jjL Q5M KM8n41 KHQLn3j.

We had heard of Dr. Robert Hatchett. A world authority on bioterrorism and toxic waste disposal, he exuded a sense of ironic good humor at the overhyped technology he heard about every day.

"Nanomachines in science fiction play the role of magic," he said with a smile. "If one needs a superflexible robot, a swarm of nanomachines fits the bill. If one needs extrasensory perception, nanomachines can do that, too. The forseeable reality however suggests that nanomachines have pico-intelligence.

"In fact the *nanomunching* machines we are planning to deploy for hazardous waste disposal will follow a trivial program of the form:

```
eat at start position
loop
    if there is something to eat to the left
        then go there and eat it
    if there is something to eat above
        then go there and eat it
    if there is something to eat to the right
        then go there and eat it
    if there is something to eat below
        then go there and eat it
end loop
```

"We abbreviate such a loop as left, up, right, down. Because the nanomuncher gets its nutrients from what it eats, it won't visit a node that it has already eaten or that another nanomuncher has partly eaten. The nanomuncher will not even go through that node. All it will do is keep following the loop on the graph it is assigned to until it can't continue.

"Note that the loop keeps its 'program counter' position after each move. So, if you start at A and go left to B, you will first try to go up from B. If not possible, then right; if not possible, then down; if not possible, then left. If none are possible, then B is called a black hole and the nanomuncher will disintegrate. The goal is to cause your nanomuncher(s) to eat at every node in a given graph before entering a black hole. If you succeed, then you have 'munched' the graph.

"You are allowed to choose the first node and to choose the order in the loop. So a loop might be right, up, left, down, for example."

"I see that both first node and loop order can make a difference," 14-year-old Liane volunteered. "Consider A —— B —— C. If you start at B, then you will surely lose no matter what your loop is. If you start at A or C you will win, regardless of the order in the loop.

"Here is an example in which the order makes a difference. Suppose you have

"If you start at C, then you will win if the loop is down, right, up, left, (or anything else that starts with down) but not if it is right, down, . . . (or anything else that starts with right)."

"Very fast on the uptake, young lady," said Hatchett. "Perhaps you can answer this one: What is the connected graph with the fewest nodes that cannot be munched no matter what the order or what the starting point?"

[Before reading on, give it a shot.]

Liane thought about this for a few seconds. "Any two- or-three node graph must constitute a single path, which can obviously be munched," she said. "So the smallest impossible connected graph must have four nodes":

"Well done again," said Hatchett. Then he posed several questions I've taken the liberty of paraphrasing here.

1. Are there any shapes having the property that even an arbitrarily large graph of that shape has a solution regardless of the start node and regardless of the order of the loop?

2. Are there graphs that are munchable with two nanomunchers (with different starting points and possibly different loop orderings) but not with one? (You can assume that if a nanomuncher arrives at a node already occupied by another, it will retreat. If two nanomunchers reach an uneaten node at the same time, the one coming from a lower node will retreat; if they both come from nodes at the same height, then the one to the left will retreat. You may not assume any other synchronization nor may you assume that different nanomunchers work at the same speed.)

How about vice versa, that is, some graph that two nanomunchers can't munch but one can? For this to be true, there must be a successful start node and loop configuration for one nanomuncher, but none for two nanomunchers. (It's not enough to find one unsuccessful configuration for two nanomunchers. All configurations must be unsuccessful. This one is open.)

"You have posed many interesting questions," Ecco said with a smile. "Is there one question that you are most concerned about?"

"Well, several actually," Hatchett responded. "You see here in this figure a graph that is a portion of a grid. Can you munch it with one nanomuncher?"

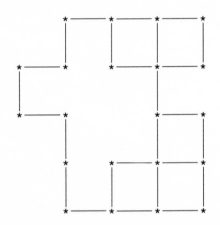

3. Can you?

4. Can you munch a 4-by-4 grid with one nanomuncher?

"Finally, we will often be faced with grids of various vertical and horizontal dimensions. Which ones are munchable by a single nanomuncher and how?"

"I take it that a grid is a two-dimensional n-by-m checkerboard configuration, right?" Liane asked.

Hatchett nodded.

5. Which n-by-m grids are munchable by a single nanomuncher?

It's open to figure out which is the largest square grid having even sides that can be munched. I know of no solution larger than four.

STRATEGIC BULLYING

"N vyl3U vf auR3U1. N Z2uSS z3Ux Zw2VSuYZ SuayY aV
xyTVUZaYuay Tf WyYzywa Y3l2a." - 9YyxyY3wR (NN) a2y LYyua
Na S3yZ d3a23U a2Yyy vSVwRZ Vz HulbuYx3u jSuwy

Will there be a war?

In schoolyards, the underworld, and international affairs, certain entities try to take valuables from others by force. They do so most happily when they have overwhelming power and the target is wealthy. This occurred in colonial wars throughout time. There is even evidence that a high proportion of deaths in prehistoric times resulted from war.

We will represent the power of each entity by a number. A coalition's power will be the sum of its members' powers. Coalition A will destroy coalition B if A challenges B and A's power is greater than B's power. If A challenges B but B's power (possibly after acquiring allies) is greater than or equal to that of A, then A will back down and no conflict will result from that challenge. So, war is far from inevitable.

When an entity is beaten, it is destroyed and its wealth (but not its power) is distributed among the winners. Its wealth is proportional to its power, however. The winners lose none of their power. Every entity plays selfishly, striving above all to avoid certain destruction but also to acquire wealth.

A configuration of power values is stable if there is a strong enough coalition to prevent any war and every member of that coalition has an interest in participating in the coalition.

The simplest stable configurations are invincible ones in which every entity can avoid being destroyed without requiring allies.

There are two kinds of invincible configurations. Can you find them without reading ahead?

X alone or X, X, for any power value X. If there is one entity, then it is clearly invincible. If there are two entities with the same power, then each is invincible (balance of power).

Let's consider more complex configurations. For example, suppose the power values are 4, 2, 1. Then 4 will fight 2 and 1 and will beat them both whether they form a coalition or not. On the other hand, if the power values are 4, 2, 1, 1, then the low-power entities can form a coalition to prevent the 4 from beating them. However, if any weaker entity is destroyed, the remaining ones will be too, so the coalition members have a strong self-interest to stay together. This configuration is stable.

Note how delicate that stability is, however. If the 4 power can convince one weaker power to attack another or, even better, simply declare itself to be neutral if 4 should attack, then 4 can conquer them all. The British used this divide-and-conquer strategy repeatedly to acquire their enormous empire. Historians may recognize that Hitler used a variant of that strategy in the fall of 1938 when he convinced the French and British to let him take the Sudetenland from Czechoslovakia, effectively rendering Czechoslovakia defenseless and thereby weakening France. For the moment though, we will assume that all entities recognize their ultimate self-interest.

WARM-UP.

Consider the power configuration 5, 4, 4. Is it stable? Can you predict what will happen? Can you predict what will happen to the configuration 5, 4, 3? Think before you read on.

SOLUTION to WARM-UP.

Configuration 5, 4, 4 is unstable, because the two 4's have every incentive to destroy 5. They will both acquire wealth and they then fall into a stable (in fact, invincible) configuration. Configuration 5,

4, 3 is stable because if any one of these is destroyed, then the weaker of the remaining ones will be destroyed and this weaker one can ally itself with any challenged entity to avoid an initial conflict.

Here are a few related quantitative questions:

1. *What is the largest set of distinct positive whole number values that is stable and whose greatest power is 21?*

2. *What if the positive whole number values need not be different?*

Until now, we've assumed that an entity will attack if it is sure of winning and ending up invincible or in a stable configuration. An entity will join a defensive coalition if it is otherwise sure to be destroyed. But life is rarely so certain: What if an attack will profit an entity but may or may not cause its destruction? To make this precise, call an entity with power X risk-averse if it will work to prevent a conflict that could *later* result in the defeat of at least one entity with power X. Call an entity with power X risk-ready if it will work to prevent a conflict that would later result in the defeat of *all* entities with power X but otherwise will allow the conflict to occur. Notice that whether entities are risk-averse or risk-ready, 5, 4, 4; 4, 4, 4; and 4, 2, 1 are both unstable whereas 5, 4, 3 and 4, 2, 1, 1 are both stable.

3. Can you show a configuration that is stable when all entities are risk-averse but not when they are all risk-ready?

So, it seems that risk-aversion would always reduce the number of conflicts. After all, risk-averse entities will not attack even when risk-ready ones will.

4. Is that true? Can you prove that a stable risk-ready configuration will always lead to a stable risk-averse configuration? Or can you show this is false by giving a counterexample?

5. What about if some entities are risk-averse and some are risk-ready? Does either have a consistent advantage in terms of survival?

This field requires a theory. Let's start with a theory of risk-averse configurations.

A coalition L dominates a configuration C if L can conquer everyone in C who is not in L. For example, 4, 2 dominates 4, 4, 2, 1 and 5 dominates 5, 3, 1.

Stability is governed by the following three rules:

  i. A configuration is stable if it is invincible.
 ii. A configuration C is unstable if it has a proper subset X such that X is stable and X dominates C.
iii. A configuration C is stable if every dominating proper subset X is unstable.

Here are examples of these definitions. The following configurations are stable because they are invincible: X; X, X. Therefore the following is unstable (because a proper subset is stable and dominating): X, $X - 1$. If X is at least 1, then X, $X + 1$, $X + 2$ is stable because the only proper subsets that are stable consist of a single entity and that entity does not dominate.

6. Can there be several stable subsets of an unstable configuration?

OPEN PROBLEM.

Does this definition capture all stable configurations for the risk-averse setting? How can it be generalized to the risk-ready setting?

Science and Form

THE STONE TOMB OF ZIMBABWE

Na Zaf 14 Y8eX43 1B f74 5z2f f7zf Bag zd4 zf X814dfB zZ3
d4Xzf8h4XB 5d44; f7zf 5ad f74 YaY4Zf Bag zd4 Zaf gZ34d Xa2W
zZ3 W4B: Bag 7zh4 e8YbXB 144Z 6dzZf43 z d4bd84h4.—qBeCzd3
Uzbge28ZeW8, 9 vzdezi N8zdB
    s74 zd4z 8e Qd44Zi827 u8XXz64.

Natasha, a reknowned full professor of archeology though still in her early thirties, entered Ecco's apartment. Apparently her identification of the most likely beam lengths of Menea had brought her fame and tenure. Success had also kindled an interest in conservation with a blend of tomb raiding.

"You've heard the story of the discovery of King Tut's tomb, Dr. Ecco, I'm sure," she said. "From 1916 to 1922, Howard Carter spent six years in unsuccessful pursuit of the tomb, enduring the unforgiving sun of the Valley of the Kings in Egypt. His funding resources nearly exhausted, he was ready to give up when his luck changed suddenly: one of his workers found a stair. The staircase led to a door. The door to an antechamber, the antechamber to . . . Carter pulled out hundreds of artifacts, many made of pure gold. Many of these treasures were preserved, but many more were destroyed by hands, pollution, and moisture.

"Recently, we have located an underground stone tomb in Zimbabwe. We believe it is airtight. But the surface air there is infinitely more moist than in the Valley of the Kings. We know that if we excavate it now, the moisture will destroy nearly everything inside, so we want to learn as much as possible by noninvasive sensing.

"Let me tell you what we know. The tomb is an enormous hexagon. There are two corridors between vertices of the hexagon (but not along the periphery). The two corridors have diamond-studded walls and are basically straight. Two corridors may even link the same two vertices (that's why I said basically straight). They may also cross one another. There are many possibilities. We just know that there are two of them and that no corridor links neighboring vertices.

"We want to determine the vertices that each corridor connects. Our remote sensing devices are scattered by the diamonds so each device consists of a radio frequency laser and a detector. The laser is placed outside an exterior line segment and shoots toward the detector which is outside another exterior line segment. The detector reports back whether the beam has passed zero, one, or two corridors."

"Diamonds?" 14-year-old Liane said. "How long do you think you will keep that secret?"

"Not long," Natasha replied. "That's why we need to move fast. I have some drawings that might help. The three rays in this hexagon aren't enough because they can't distinguish $(0,3)$ from $(1,4)$ or $(2,5)$:

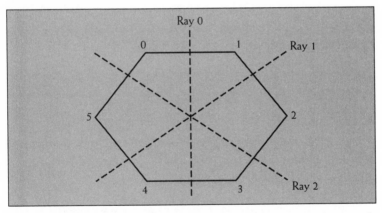

Figure 23a These three rays are not enough, because, for example, the pairs of lines/corridors $(0,2)$ and $(0,3)$ together intersect the same rays as $(0,2)$ and $(1,4)$. Both pairs intersect ray 0 twice, ray 1 twice, and ray 2 once. The basic problem is that $(0,3)$ and $(1,4)$ intersect the same rays (all three).

"Similarly, the three rays in this hexagon are not enough because they give no way to distinguish corridors $(0,2)$ from $(2,4)$:

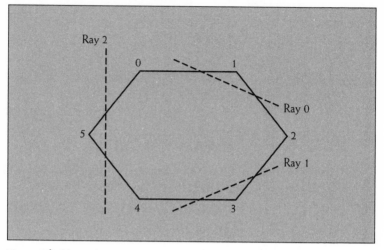

Figure 23b These three rays are also not enough, for the simple reason that they cannot distinguish between (0,2) and (2,4), for example.

"Your job is to set up as few rays (laser-detector pairs) as possible, so that no matter which corridors are present, you will find out which pairs of vertices they connect.

1. How many rays do you need if you must set up all the rays before turning on any laser and you must be able to identify the locations of the corridors for sure?

Liane handed Natasha her six-ray solution. Natasha studied it. "Yes, I see. Well, if that's the best you can do, I thank you for your efforts once again." She was remarkably perfunctory I thought, considering that Liane had made her career.

2. Can you generalize this solution to higher n-gons (n > 6)? [hard]

"Nice job," Ecco said to Liane after Natasha left. "But you may be able to do better if you change the rules a little. Suppose you could set up one laser-detector pair, detect the ray, record the result, then set it up somewhere else, detect the ray, record the result, and so on.

"Here's why I think you might do better. Each ray can give three responses, so the total information of four rays is $3^4 = 81$ possible responses. And there are fewer than 81 alternative layouts of the corridors, as you can see from the following argument. There are six possible corridors that go between vertex x and $x + 2$ mod 6. There are three possible corridors that go between vertex x and $x + 3$ mod 6. So there are nine possibilities for each corridor. Since there are two corridors, there are 81 possible configurations of the two corridors, but those 81 distinguish between the first corridor and second corridor which is unnecessary, so this double-counts most combinations. Therefore four rays should be more than sufficient. Can you do that well?"

3. *How many setups would you need? The laser beams still go from a segment of the hexagon to another segment.*

PROTEIN CHIME

"a Q8GN hiE O8RfNM. a'GN dFDE OiFhM j0,000 H8JD EQ8E
Hih'E HiCe." - 1Qig8D SfG8 WMRDih (jqmp-jrlj)
vhN i0 EQNDN QRhED g8J 9N 8 fRN 9FE D8JD Di.

In the 1950s Jacques Monod and François Jacob of the Pasteur Institute showed that certain regulator proteins in *E. Coli* bacteria can repress the production of other proteins. For the sake of this puzzle, let's use the notation $X \rightarrow Y$ to mean that protein X represses protein Y. If X goes up (that is, the protein appears), then Y will go down (the protein disappears) after a short time (say, one second). If X goes down and no other repressor for Y is up, then Y will go up one second after X goes down.

Now consider three proteins A, B, and C such that $A \rightarrow B \rightarrow C \rightarrow A$. If A goes up, B goes down one second later. After another second, C goes up, and after one more second A goes back down. Then the pattern continues: B up, C down, A up, B down, and so on. We call this a circuit of proteins: A, B, and C periodically appear and disappear, acting like a biochemical clock. Researchers Michael Elowitz and his advisor Stanislas Leibler from Princeton have actually constructed such a clock. They used an odd number of proteins in each circuit because any even-numbered circuit could be stable: half the proteins would stay up all the time and half would stay down.

Now let us assume there are eight circuits, labeled A through H, each containing three, five, seven, or nine proteins:

A proteins: $A1 \rightarrow A2 \rightarrow A3 \rightarrow A1$

B proteins: $B1 \rightarrow B2 \rightarrow B3 \rightarrow B1$

C proteins: $C1 \rightarrow C2 \rightarrow C3 \rightarrow C4 \rightarrow C5 \rightarrow C1$

D proteins: $D1 \rightarrow D2 \rightarrow D3 \rightarrow D4 \rightarrow D5 \rightarrow D1$

E proteins: E1 → E2 → E3 → E4 → E5 → E6 → E7 → E1

F proteins: F1 → F2 → F3 → F4 → F5 → F6 → F7 → F1

G proteins: G1 → G2 → G3 → G4 → G5 → G6 → G7 →> G8 → G9 → G1

H proteins: H1 → H2 → H3 → H4 → H5 → H6 → H7 → H8 → H9 → H1

A1 → T

B1 → T

C1 → T

D1 → T

E1 → T

F1 → T

G1 → T

H1 → T

No protein in any circuit has an effect on proteins in different circuits, but one protein in each circuit represses T, the special chiming protein that rings when it goes up. If any of these eight repressor proteins (labeled A1 through H1) goes up, T will go down one second later. And T will not go back up until one second after all eight of the repressor proteins are down.

Further, you may force a particular protein in a circuit up or down. For example, to start the C circuit, you may choose to keep C4 forced up for five seconds. One second after the forcing begins, C5 goes down. In the succeeding seconds, C1 goes up, C2 goes down, and C3 goes up. But C4 does not go down at the end of five seconds, because it is still being forced up. C3 does not push C4 down until one second after the forcing stops (that is, six seconds after the start). Then the cycle resumes without interruption: in the seventh second, C5 goes up and in the eighth second C1 goes down. C1 goes up again in the thirteenth second, down again in the eighteenth, and so on. Note that if you force no elements in a circuit, roughly half of the elements will be up and half will be down, you just won't control which ones.

Can you create a "protein chime" that will ring every 70 seconds? Note that you may not need to build all eight circuits. (The excluded circuits have no effect on T.) Your challenge is to determine which circuits must be activated and how to start each one: which protein in the circuit must be forced up and for how long. You may force a protein up during the first 10 seconds; after that, the chime must operate on its own.

LOST HIKER

"GyF zH 5JG 3zCF HGGwKyI D65 4z 7G3L Ewz4G 5z 5z8y"
qz64G CF.
    eJG CFF3G44 K4 8K5JKy HK7G DwzEv4 zH HKH5GGy5J
453GG5 (wKG).

A hiker has been seriously hurt somewhere in a 10-mile-square area of wilderness. He sends a distress signal having a range of two miles (see the circles of Figure 24). When your search party is within range of his signal, your directional finder will lead you directly to him, so your task is to *guarantee* to come within range of that signal as quickly as possible. Assume you can start from any edge of the square and travel continuously a mile every 10 minutes.

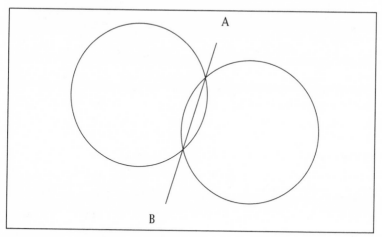

Figure 24

WARM-UP.

Suppose the rectangle, instead of being 10 miles square, were 4 miles wide by 25 miles long. How long would it take you then? Try before you read on.

SOLUTION to WARM-UP.

Well, you certainly could do it in 250 minutes by simply driving down the middle, starting at the narrow end of the rectangle (see Figure 25). However, you may notice that as the jeep drives to the far end of the rectangle, it's quite inefficient at covering area at the end. You can exploit this to find a better solution. Please give it a try before you read on.

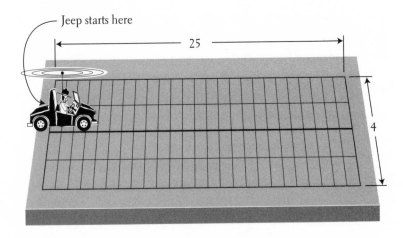

Figure 25 A 250 minute solution if you have a rectangle 4 miles wide and 25 miles long. There is a better one.

BETTER SOLUTION to WARM-UP.

A better solution is to drive the jeep 23 miles down, at which point, the jeep is $2\sqrt{2}$ miles from the two corners, so it goes $2(\sqrt{2} - 1)$ miles toward one corner and then directly across until it is 2 miles from the other corner. The total distance is approximately 24.58 miles, but this could be improved further by departing from the line earlier (see Figure 26). This suggestion came from Dean Ballard. Rudolf de Heus managed to bring this down to 24.535. See his website at www.xs4all.nl/~rheus/Hiker/SciAmSept2003HikerWarmUp.htm.

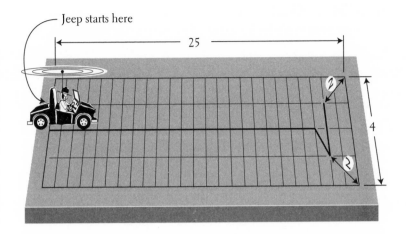

Jeep starts here

25

4

Figure 26 A sub-250-minute solution if you have a rectangle 4 miles wide and 25 miles long. The drawing is off scale, but the idea is to veer off toward one corner and then cross over to cover the other.

To give a best bound of what can be done with the square, it is good to know which shape that is 100 square miles in area gives us the smallest possible time.

1. *Can you find the best planar shape and the best route on that shape?*

2. *What is the best route you can find through the 10-by-10 square?*

3. *What if, to conserve energy, the distress signal goes on for one second every 10 minutes? What would be the best route then?*

NOAH'S ARC

"6Rt, SOuqVu VqYu Pu vURP bRXU vROORZuUV!" - 1XPSuU
KWysNuU

mxqW yV yW qrRXW RXU VSusyuV WxqW PqNuV
MuqORXVb,VR PXsx PRUu SRZuUvXO qQ uPRWyRQ WxqQ
sRQWuQWPuQW?jxqW PqNuV uQYb qQt UuVuQWPuQW qQ uqVb
yQtXsuPuQW WR YyROuQsu?GWxuU VSusyuV vywxW RYuU PqWuV
qQt vRRt qQt WuUUyWRUb.mu vywxW RYuU SqUWysXOqU rUqQtV
Rv PRQRWxuyVP RU WxuyVP, rXW PRVWRv qOO Zu vywxW ZxuQ
Zu qUu xXPyOyqWut.mxb yV WxyV VR xqUt WR XQtuUVWqQt?
jxu qXWxRU ZyOO PuuW bRX WxuUu qW uOuYuQ.

Dr. Windswift was back. "We are cloning, er, higher mammals on our floating laboratory in the South Pacific outside all territorial limits. The ship is called *Noah's Arc,* but you must excuse the poetic license. We don't need two parents for each species, just the sex of interest. We give our mammals certain genetic advantages by inserting specific DNA sent to us by suppliers. Those suppliers are also our competitors, however, so we want to verify the DNA they send.

"To do that, we will use a microarray. In each cell of the microarray, we will place many identical short sequences of DNA, called strands. If we expose the supplied DNA sequence (S) to the chip, any cell in which each strand is a consecutive subsequence of S will light up; no others will light up. (Technically, we put the reverse complement of those strands on the microarrays, but the mathematics doesn't change, so let's forget that.)

"For example, if the sequence S is AGGTCACGTGG, then a cell whose strands are CACG will light up but one consisting of CTCG strands will not. Similarly, a GG cell will light up but TT will not. Also, GG will not light up with greater intensity just because it appears twice.

"In general, if I give you a sequence, I would like to know what to put in the cells of my microarray. I would like the maximum length strand to be as short as possible and within that length, I'd like to use as few strands as possible.

"Oh, I almost forgot. Strands that match the leftmost end of the DNA sequence will light up with a special color, so you'll know which ones those are. If a cell's strands are in the leftmost end and somewhere else, then the cell will light up with both colors. Also, by independent means, we can tell you the individual totals of A's, C's, T's, and G's."

WARM-UP.

"What is the smallest number of strands needed for some arbitrarily long sequence?" Windswift asks.

SOLUTION to WARM-UP.

"That's too easy," Liane answered with a chuckle. "Any sequence consisting entirely of a single nucleotide requires no strands at all."

"You're right," Windswift answered. "It was a trick question."
Now try these:

1. What is the shortest sequence S you can imagine such that even if strands can be six characters long, you won't be able to verify S?

2. If S = ACGAC, then two strands of length 2 are enough. Can you see how?

3. What if we have AC repeated N times?

4. Here is the sequence we want to verify:

TCACTCGGCTCTCGCACACGGAGATAGCTC.

What is the smallest size and the smallest number of strands of that size that would verify this sequence?

LIQUID SWITCHBOARD

"TiN FRO hAEF ShBADF9iF FRSiQ FA DOhOhLOD SE, iA
h9FFOD IR9F 9iKLANK FOggE KAG,SF SE iOHOD, OHOD
GiB9FDSAFSM AD Gi-ThODSM9i FA CGOEFSAi 9iK PGMfSi' FRSiQ
Si 9NOhAMD9MK," 1FOHO X9DgO, ESiQOD.
    2RO 9GFRAD ISgg QSHO KAG hADO SiPADh9FSAi.

Imagine a collection of pipes arranged in a circular fashion, labeled
alphabetically A, B, C, D, E. Each letter stands for the color of the
water at the top of the pipe: amber, blue, crimson, diamond, and
emerald, respectively.

At three places along the vertical axis of each pipe, there is a potential switch with each neighboring pipe (see Figure 27). If a switch point between pipes X and Y is set, then whichever color liquid is flowing in X above that switch point flows down the Y pipe after the switch point and vice versa. At a given height of pipe X, at most one switch point may be set. So X can swap liquids with no pipe, the pipe to its right or the pipe to its left at that height.

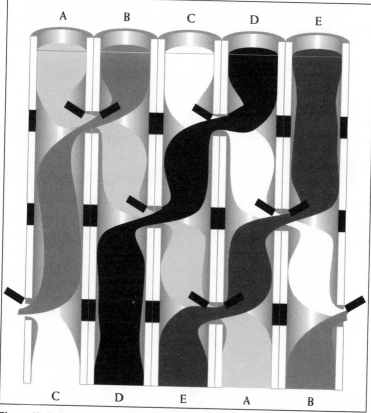

Figure 27 A through E are arranged in a circular pattern with three switch opportunities along every pipe. A pipe can switch with at most one neighbor at each opportunity.

WARM-UP.

How can you arrange the switch points so that A, B, C, D, E becomes C, D, E, A, B at the bottom? That is, crimson is at the bottom of the amber pipe, diamond at the bottom of the blue pipe, and so on.

SOLUTION to WARM-UP.

At the top switch level, switch between A and B and between C and D. This gives B, A, D, C, E. Then switch A and D and switch C and E. This gives B, D, A, E, C. Finally, switch B with C and A with E. This gives C, D, E, A, B as desired.

This rearrangement (aka permutation) is so far from the original that one might think that any rearrangement of these five colors is possible using all these switches.

Is that true? If not, which ones are not possible and how many more levels would you need to make every rearrangement possible?

Call a configurations of P pipes and L levels of switches "permutable" if all possible permutations of the P colors can be arranged by some setting of the switches. Adding levels of switches keeps a permutable configuration still permutable, so the question of interest is how small L can be for a given P to guarantee permutability. I don't know the answer.

PRIME SQUARES

"4 mL MNS XNTMs qMNTst SN 8MNV qUqQXStuMs." -
CRomQ Ku9pq (Zgdc-Zh00)
   Htq qUqMS Sm8qR 09moq mrSqQ SVN StNTRmMp mMp ruUq
nqsuMR.

Prime numbers hold a strange fascination for mathematicians of all
ages. Children grasp the idea quickly and soon recognize that primes
are much more dense among small numbers than among big numbers. On the other hand, professional mathematicians have not yet
proved nor disproved the conjecture that every even number greater
than four is the sum of two primes, a conjecture made by Goldbach
and Euler in 1742.

   This puzzle asks you to find prime squares: a square grid, each of
whose cells contains a digit (between 0 and 9) that form primes when
extended to rows and columns. No two rows and no two columns are
allowed to be the same. An ambidextrous prime square is one which
also contains primes when the rows are read from right to left. An
omnidextrous prime square is an ambidextrous prime square whose
columns are primes when read bottom to top and whose diagonals
are also primes. One other constraint on prime squares: they cannot
begin with a 0.

   Here is an example of an ambidextrous prime 3-square:

7	6	9
9	5	3
7	9	7

Because 769, 953, and 797 are all primes as are their columns 797,
659, and 937, this is a prime square. Because the row reversals 967,

359, and 797 are also primes, this is an ambidextrous prime square. One reason it is not omnidextrous is that 956 is even.

WARM-UP.

Try to find a prime 4-square that uses nine distinct digits.

SOLUTION to WARM-UP.

3	2	5	3
8	4	1	9
8	2	6	3
9	3	7	1

1. Can you find an omnidextrous prime 3-square that uses as few distinct digits as possible?

2. How about a prime 5-square that uses all 10 digits?

It is a research problem to find the set of numbers *n* having prime *n*-squares (whether plain, ambidextrous, or omnidextrous).

Finally, see if you can win a game. It's called tictacprime. You play it on a 3-by-3 grid. A move consists of placing a single digit in an empty cell. The two players alternate moves except the second player gets the ninth move. When a player's digit completes a segment of three digits in a row, then the number of three-digit primes that include the square where the digit was placed (going in any direction) yields a point to that player. The object is to get the most points possible. Is there a winning strategy for either player? That's for you to play out. It might help you to have a list of three-digit primes:

193	307	421	547	659	797	929
191	293	419	541	653	787	919
181	283	409	523	647	773	911
179	281	401	521	643	769	907
173	277	397	509	641	761	887
167	271	389	503	631	757	883
163	269	383	499	619	751	881
157	263	379	491	617	743	877
151	257	373	487	613	739	863
149	251	367	479	607	733	859	997
139	241	359	467	601	727	857	991
137	239	353	463	599	719	853	983
131	233	349	461	593	709	839	977
127	229	347	457	587	701	829	971
113	227	337	449	577	691	827	967
109	223	331	443	571	683	823	953
107	211	317	439	569	677	821	947
103	199	313	433	563	673	811	941
101	197	311	431	557	661	809	937

Figure 28 Three-digit primes between 100 and 999.

REPELLANOIDS

"uXd ywW 31c Vdy4 2dac41a f5c4 w T5Wz fXaz wWz w 3dW
c4wW hXd ywW f5c4 w T5Wz fXaz wUXW1." —6U 8wYXW1
p41 le1Wc cwT1b YUwy1 x12Xa1 cfX c4XdbwWz wWz c1W.

A virus called a repellanoid has been implicated in a host of diseases.
Biochemists have determined that it forms a cylinder but don't know
the length of its circumference. Here is what they do know. There
are five kinds of strands that they refer to by colors: aqua strands of
length 4; blue of length 5; crimson, 6; daisy, 7; and green, 8. Each
ring of the cylinder consists of strands laid out end to end.

The biochemists have identified the following constraints: taking the vertical direction to be along the length of the cylinder, if X
is a ring and Y is the ring immediately above X, the ring two above
X, or the ring three above X, then any vertical line between X and Y
must touch a different color on X than on Y. That is, like repels like
up to a distance of 3. All the rings must be of the same circumference.

WARM-UP.

Suppose that like repels like at a distance of 2 only. But the only
strand types are aqua, blue, crimson, and daisy. What would be the
smallest circumference then?

SOLUTION to WARM-UP.

Twelve. One ring could be aqua, aqua, aqua; its neighbor, crimson,
crimson; and its neighbor, blue, daisy. Then this pattern could be
repeated.

1. What is the smallest circumference of the cylinder in the full problem?

*2. A second research group has discovered a second virus called a mini-repellanoid.
Here is what they know: every strand has length 2 or greater, there are five different strand sizes, the circumference is 7, and like repels like at a distance of 3. How
could this be?*

Combinatorics

SAFECRACKING

"Xifo J bn xpslAoh po b qspcmfn J ofwfs uiAol bcpvu cfbvuz. J pomz uiAol bcpvu ipx up tpmwf uif qspcmfn. Cvu xifo J ibwf gAoAtife, Ag uif tpmvuApo At opu cfbvuAgvm, J lopx Au At xspoh." - CvdlnAotufs Gvmmfs (29L6-2L94)

Uif xfbuifs nbz ps nbz opu cf xbsn.

Imagine that you are a thief (for a good cause, naturally). At great personal and societal risk, you must steal some papers from an evil-doer. To accomplish this, you must find the combination of a safe having 10 switches, each of which has three values: low, middle, and high. To open the safe, you must set the switches to an opening combination and press the lever. To escape detection, you must do this as quickly as possible.

There are over 59,000 (3^{10}) possible combinations. Fortunately for you, over 6,000 (3^8) combinations of these switches will open the safe. The rule to open is simple: there is a certain pair of switches—you don't know which pair and they need not be adjacent—and a certain combination of the values of that pair—you don't know which combination—that opens the lock. The other eight switch settings are irrelevant. Now, you want to guarantee to open the safe in as few lever pulls as you can.

You realize that if you try random settings, then you have a 1 in 9 ($3^8/3^{10}$) chance of finding a good setting. Your problem, however, is to *guarantee* to find a solution within fewer than 20 switch settings.

How can you find the combination?

WARM-UP.

Suppose there were just four switches, each with three values, but again there is some pair of switches and some combination of values of that pair that would open the lock. How many lever pulls would you need to guarantee to open the lock then?

SOLUTION to WARM-UP.

Well, nine is clearly a lower bound, because even if you knew the pair of switches to choose, you'd have to discover the pair of values of those switches. Remarkably, nine are enough (here A means low, B means middle, and C means high):

Number	S1	S2	S3	S4
1	A	A	A	A
2	B	C	C	B
3	C	B	B	C
4	A	B	C	C
5	B	A	B	A
6	C	C	A	B
7	A	C	B	B
8	B	B	A	C
9	C	A	C	A

This was first pointed out by Rob Pratt of the University of North Carolina and independently by Michael Bookbinder, Brett Stevens, Ivan Rezanka, Wim Caspers, and Joe Binsack.

OPEN PROBLEM.

Suppose the cost, instead of being the number of lever pushes, were the total number of switches that must be moved. What would be a minimum cost solution under that metric and assuming that the lever does not reset all the switches upon each pull. I don't know the answer.

HARD TO SURPRISE

"fT mEK VOjSD'J UEJ ODmJVWDU DWQS JE IOm OPEKJ ODmPERm, QECS IWJ DSIJ JE CS." - XBWQS 4EEISjSBJ xEDUkEHJV (ovvr-owv0)
6VS SjSDJ JOASI FBOQS PSJkSSD JSD OC ODR JkSBjS DEED.

People who repeat themselves are really annoying. Really annoying. Really. Annoying.

In this puzzle, the goal is to study the mathematics of repetition. Let's consider a sequence of symbols (each one representing a word or perhaps several words). We will call a sequence "surprising" if, for every pair of symbols X and Y and any distance D, there is at most one position in the sequence where X precedes Y by distance D. For example, the top two sentences have the same distance between "really" and "annoying."

Here are three more symbolic examples: AAB is surprising as is AABA, but AABB is not, because there are two instances when A is followed by B two symbols away (by distance 2).

WARM-UP.

Show that the following sequence composed from the symbols A, B, and C is not surprising: BCBABCC. Find a surprising sequence using A, B, and C that is at least seven symbols long.

SOLUTION to WARM-UP.

BCBABCC is not surprising because the letter B precedes a second B by distance 2 two times. Here is a surprising sequence of length 7: BACCBCA.

Here are three much harder challenges:

1. Find surprising sequences that are as long as possible and that can be composed from 5, 10, or 26 distinct symbols. For convenience, use the letters of the alphabet as symbols. One hint is that the length doesn't increase very fast, and I believe that it is always under 100 (more on this conjecture in the solution).

The notion of surprise above is called 2-surprising because only pairs are involved. We could also define 3-surprising to mean that for any triplet of symbols X, Y, and Z and distances $D1$ and $D2$, there is at most one position in the sequence where the first symbol (X) precedes the second (Y) by distance $D1$ and the Y precedes the third (Z) by $D2$.

2. What is the longest 3-surprising sequence you can find composed of the first five letters of the alphabet?

3. I know of no simple rule that will give the longest possible k-surprising sequences composed from sets of n symbols for any k and n. Can you find a pretty theory?

TANKTOPS AND SUNGLASSES

"HTnRVvNt nRTvST pUTr unPPY vN 8rnNS WvTu onRrsrrTOR
qRrSSrq UP vN SvL9." ErRSONnLS nq.
5T vS vN Tur svRST qrpnqr Os Tur MvLLrNvUM.

A group of mathematicians who happen to be teenage girls decides to form a fashion gang. The rules of the gang are that on each day, each girl must wear a tanktop that is either blue or black; sunglass rims that are black or brown; capris that are either black, red, white, or pink; and lipstick that is either pink, red, or brown. However, each pair of girls must differ in at least two of these items; for example, if they have the same color tanktop and lipstick, then they must differ in their choice of sunglass rims and capris. Differing in more items is quite acceptable.

The girls want to determine how large or small a gang they can form. So they have two questions:

1. What is the largest number of girls who could be in this gang and what might each girl wear in that case?

2. What is the minimum number of girls who could be in this gang to satisfy the difference constraint but such that adding one girl would violate the constraint? Again show what each girl might wear in that case.

WARM-UP.

Suppose there are just three items of clothing and the choices are all binary: a tanktop that is either blue or black; sunglass rims that are black or brown; and capris that are either black or red. Try to find gang dressings as small as two and as large as four such that adding a girl is impossible in either case. Think before you read on.

SOLUTION to WARM-UP.

For a gang of two: blue tanktop, black sunglass rims, black capris; black tanktop, brown sunglass rims, red capris

Here is a gang of four: blue tanktop, black sunglass rims, black capris; blue tanktop, brown sunglass rims, red capris; black tanktop, black sunglass rims, red capris; black tanktop, brown sunglass rims, black capris.

RENDEZVOUS OF THE STARS

The encrypted preamble to each puzzle both rants and informs. Combine the hints to find times and locations. I hope to see you later this decade.

Solutions

SHIFTY WITNESSES **SOLUTION**

Can you tell which suspects have drugs given only that the total number of lies among all witnesses is eight or nine and most of the lies claim "has no drugs" when the truth is "has drugs"?

There must be at least five lies that assert innocence ("no drugs") when they should assert guilt. We'll call those that assert innocence falsely FI and those that assert guilt falsely FG. Because the minimum number of lies among the disagreeing groups is six, all groups that are unanimous must be telling the truth. So we focus on the disagreeing votes:

Suspect 3: three vote "has no drugs" and two vote "has drugs";
Suspect 5: four vote "has drugs" and one votes "has no drugs";
Suspect 7: three vote "has drugs" and two vote "has no drugs";
Suspect 10: four vote "has no drugs" and one votes "has drugs."

First, we see that there cannot be exactly eight lies because that can happen only if the witnesses vote as often for drugs as against them (only in the 3, 1, 3, 1 case). So there must be exactly nine lies.

The only way there could be exactly nine lies is if the majority pertaining to either suspect 5 or suspect 10 lie and neither majority pertaining to suspect 3 or suspect 7 lies. This gives us two possible underlying true situations:

A. 3, no drugs; 7, drugs; 5, drugs; 10, drugs
B. 3, no drugs; 7, drugs; 5, no drugs; 10, no drugs

In situation A, there are two FIs for suspect 7, one for suspect 5, and four for suspect 10. That gives seven FIs. In situation B, there are two FIs for suspect 7, but no others. So, situation A is the only possible scenario given that the FIs outnumber the FGs.

Because the unanimous groups all tell the truth, the police should arrest suspects 1, 4, 5, 7, 8, and 10.

TERMINATING LEAKS **SOLUTION**

1. Can you help the governor guarantee to find the leakers using 25 tidbits and two leaks in all, assuming he follows the given strategy?

The governor can find the three leakers by using 25 or fewer tidbits and suffering at most two leaks by using the following construction due to New York University/Courant Institute mathematician Joel Spencer.

Label the advisors A0 through A8 (A0 .. A8) and form the foursomes according to the following protocol:

A7, $x, x + 1, x + 3$
A8, $x, x + 2, x + 3$
$x, x + 1, x + 2, x + 4$
A7, A8, A0, A1
A7, A8, A2, A3
A7, A8, A4, A5
A7, A8, A6, A0

where x ranges over 0, 1, 2, 3, 4, 5, 6 in each of the first three lines, and the "+" operation is modulo 7. Number j represents advisor Aj, for example, 3 represents A3. (Modulo 7 works like base 7 except that each operation retains only the lowest order digit. So, $3 + 3 = 6$, but $4 + 3 = 0$ [instead of 10 in base 7] and $6 + 4 = 3$.)

The construction yields the following foursomes:

A7, A0, A1, A3
A7, A1, A2, A4
A7, A2, A3, A5
A7, A3, A4, A6
A7, A4, A5, A0
A7, A5, A6, A1
A7, A6, A0, A2
A8, A0, A2, A3
A8, A1, A3, A4

A8, A2, A4, A5

A8, A3, A5, A6

A8, A4, A6, A0

A8, A5, A0, A1

A8, A6, A1, A2

A0, A1, A2, A4

A1, A2, A3, A5

A2, A3, A4, A6

A3, A4, A5, A0

A4, A5, A6, A1

A5, A6, A0, A2

A6, A0, A1, A3

A7, A8, A0, A1

A7, A8, A2, A3

A7, A8, A4, A5

A7, A8, A6, A0

Together, these foursomes include every triplet that can be formed from the nine advisors. The first cluster, for example, covers all possible triples having A7 and two members of A0 through A6 (because those two must differ by either 1, 2, or 3 modulo 7; for example, 5 and 1 differ by 3 because $5 + 3 = 1$ modulo 7). The fourth cluster covers all possible triples having both A7 and A8.

The governor would apply the following method to find the culpable triplet. Once a tidbit given to a quartet produces a leak, he would determine which of the four triplet subsets of the quartet cannot yet be cleared of guilt. If more than one triplet remains suspect, the governor would give tidbits to all but one of those triplets. Here he is taking advantage of the fact that at most three people are leakers. This strategy will identify the leaking triplet directly or by elimination. Twenty-five tidbits are the most that are needed using some mixture of quartet and triplet tests.

For example, if the governor finds a leak in the first quartet, he will need three more tests on triplets, giving a total of four tidbits. The number of tidbits required differs depending on which quartet yields the leak. To find how many would be needed if the first leak occurred at a given quartet, see the table below.

Leak Discovered	Number of Tidbits Required
1	4
2	5
3	6
4	7
5	8
6	9
7	10
8	11
9	12
10	13
11	14
12	15
13	16
14	17
15	17
16	18
17	19
18	20
19	21
20	22
21	23
22	23
23	24
24	25
25	25

For example, suppose the 15th tidbit is the first to produce a leak. This is the tidbit given to the first quartet in the third cluster above—A0, A1, A2, A4. If that foursome had come first, we would have had to test three out of the four triplet subsets within it. But, in this case, we know from having given a tidbit to the second quartet in the first cluster (A7, A1, A2, A4) that A1, A2, A4 will not yield a leak; hence, only three triplets remain suspect. Therefore, thanks to elimination, only two more tidbits need to be tested.

Carlos Gerardo Arroyo has observed that it is possible to find the triplet without any leaks. If after the 24th tidbit no leaks have been found, then the answer must be A6, A7, and A8. The 25th test is not actually needed. (If the 24th tidbit does discover a leak, then a further test is needed to distinguish between A4, A7, A8 and A5, A7, A8.)

2. Can you find the precise triplet using fewer tidbits in this case, perhaps tolerating more leaks?

If the governor accepted more leaks, he could find the guilty triplet by using no more than eight tidbits. He would start by telling a tidbit to eight advisors A0 through A7. (We denote this by A0 .. A7.) If a leak occurred, he would know that the octet contained within it all three of the loose-lipped advisors required for a newsbreak and that A8 must therefore be innocent. On the other hand, if no leak occurred, the governor would know by elimination that A8 was among the guilty and should be told all future tidbits. The governor would then tell a tidbit to A0 .. A6 and any known leaker. If that produced no leak, he would know that A7 (the only excluded advisor) was a leaker. Otherwise, he would test A0 .. A5 and any known leakers. He would continue like this for A0 .. A4, A0 .. A3, and A0 .. A2, each time accompanied by known leakers. If at any point the governor found three known leakers, he'd be done.

But after testing A0 .. A2, he might have identified only zero, one, or two leakers. Let's treat each possibility in turn. If he has discovered no leakers, then A0 .. A2 must all be leakers. If he has identified one leaker, then he can try A0, A1, and the known leaker (L), and if he still gets no leak, he can try A0, A2, L. If neither trio produces a leak, he knows that A1, A2, and L are the guilty parties. If he has identified two leakers after testing A0 .. A2, he would test them in combination with A0 first and then, if he gets no leak, with A1. After two tests, the requisite three leakers will be known, making a grand total of eight tests.

Here is one possible execution of this protocol:

Tell A0, A1, A2, A3, A4, A5, A6, A7

there is a leak, so A8 is not a leaker.

Tell A0, A1, A2, A3, A4, A5, A6

there is a leak, so A7 is not a leaker.

Tell A0, A1, A2, A3, A4, A5

there is no leak, so A6 becomes a known leaker.

Tell A0, A1, A2, A3, A4, A6

there is a leak, so A5 is not a leaker.

Tell A0, A1, A2, A3, A6

there is a leak, so A4 is also a known leaker.

Tell A0, A1, A2, A4, A6

there is a leak, so A3 is not a leaker.

At this point, there are two known leakers (A4 and A6), so one of A0, A1, A2 must be a leaker. We will test two of these in turn. Start with A0, A4, A6. There is no leak. So A0 is not a leaker. Now test A1, A4, A6. There is no leak. So A1 is not a leaker, but A2 is. This gives the set of leakers: A2, A4, A6.

3. *What if the governor were willing to ask more than four people, but still wants only two leaks?*

The following solution uses more than foursomes and requires at most 13 tidbits and two leaks. The number of threesomes that need to be checked is listed for each grouping (given that the previous tests resulted in no leaks).

A0,	A1,	A2,	A3,	A4		(10)
A0,	A1,	A2,	A3,	A4,	A5	(10)
A0,	A1,	A2,	A3,	A4,	A6	(10)
A0,	A1,	A2,	A3,	A4,	A7	(10)
A0,	A1,	A2,	A5,	A6,	A7	(9)
A3,	A4,	A5,	A6,	A8		(8)
A0,	A1,	A3,	A5,	A6,	A8	(7)
A0,	A1,	A2,	A4,	A5,	A8	(6)
A3,	A4,	A5,	A6,	A7		(5)
A3,	A4,	A5,	A6,	A7,	A8	(4)
A2,	A3,	A6,	A7,	A8		(3)
Remaining threesomes:						(2)

(84) Average tidbits = 9.55

Process of elimination enables us to eliminate the last test. This clever solution is due to Doug Bell.

NOTE.

James P. Ferry of the Center for Simulation of Advanced Rockets at the University of Illinois has a solution for the general problem of n advisors, k leakers, and L leaks allowed by carefully counting the number of possible decision trees. Each node of the decision tree is a collection of advisors and the branches are to the right (leak) or to the left (no leak). There can be only L branches to the right if L leaks are allowed.

COMPETITIVE CLAIRVOYANCE
SOLUTION

1. *Can you find a strategy that will guarantee a regret ratio (in this case, the clairvoyant's winnings divided by your winnings) that is no more than 1.8?*

When the possible offers are $1 and $5, the best strategy is to take the first 50 offers no matter what and then take only $5 offers. If the ticket exchanger offers $1 for all tickets and then halts the trading, your regret ratio would be 1.8 ($90 divided by $50). If the man initially offers $1 then switches to $5, the regret ratio would still be 1.8 ($450 divided by $250).

2. *What would be your strategy if the two possible offers were $1 and $1 million? Does the regret ratio improve or worsen?*

Take the first 45 offers no matter what and then wait for $1 million offers. This strategy yields a regret ratio of 2.

3. *Can you find a general solution for any pair of offer values?*

Suppose the minimum value is 1 and the maximum is M. You have T tickets. The general strategy is to take the first x offers no matter what they are and then take only $M offers after that. If all offers are $1, the clairvoyant oracle will get T, so the regret ratio will be T/x. If there are at least T $M offers, but they come after the first x have been given out, then the clairvoyant will get MT, making the regret ratio $MT/(x + M(T - x))$.

Our challenge is to maximize the minimum of these two. This happens when we set them equal:

$T/x = MT/(x + M(T - x))$ or $1/x = M/(x + M(T - x))$

or $MT - (M - 1)x = Mx$

or $MT = (2M - 1)x$

So, $x = MT/(2M - 1)$.

When M is $5 and T is 90, $x = 5T/9$, and $x = 50$.

When M is $1 million, $x = 90,000,000/1,999,999$, or about 45.

So, you take the first 45 and then wait for the millions. Your regret ratio will be 2.

4. *What if there were three or more offer values, say* u, v, w *in increasing order?*

The general strategy generalizes the one above: you take the first x no matter what. Then you take y if they are v or more and then wait for the w values. If all offers are u, the clairvoyant oracle will get value Tu, so the regret ratio will be T/x. If there are no w offers, but

(147)

Tv offers that come after the first x, then the clairvoyant oracle will receive value Tv and the regret ratio will be $vT/(ux + v(T - x))$.

If there are at least $T w$ offers, but they come after the first xu values and yv values have been given out, then the clairvoyant will get Tw, making the regret ratio $Tw/(ux + vy + w(T - (x + y)))$. The goal is to maximize the minimum of these. Again, set them equal and solve. Generalizing to k should be done with machine help.

GOING SOUTH **SOLUTION**

1. Can you find a testing method that will give you a probability of 3/4 or better that you can take five flares that will all work and that requires testing as few flares as possible?

If you want five good flares and there are three bad ones in the bad pile, then you can get them with a probability of 3/4 with the following strategy. Test one flare from one pile, call it pile A. If the flare is good, take the remaining ones from A. Otherwise take five from B. If A is good (probability 1/2), then you will correctly choose five from A. If A is bad (probability 1/2), then you will discover this with probability 3/6 by choosing a single flare. So this gives an additional probability of 1/4. Because A cannot be both good and bad, the two probabilities add to 3/4.

One difficulty with this is that you know that your chances of having five good flares are high, but most of the time you don't know for sure whether the flares will be good until you try them. In fact, you know for sure only 1/4 of the time (when the tested flare is bad). To increase the probability of being sure when you go to the Antarctic, you can test one flare from each pile. This will reveal the bad pile with probability 1/2, thus increasing your certainty. Overall, you chance of having five good flares remains at 3/4, but half the time, you can pack off to the arctic with a confident smile on your face. This advantage was pointed out by Lothar Koch of Hamburg.

2. *What would be your testing strategy if you needed five good flares but there were four bad flares in the bad pile? What would your probability of success be?*

You can use the same strategy as in problem 1. Again, pile A is good with probability 1/2. If pile A is bad, you will find out with probability 4/6 and choose pile B. So, your chance of winning is 1/2 + (1/2 × 4/6) = 1/2 + 1/3 = 5/6. So four bad flares actually help your chances of success.

3. *Returning to the case where there are three bad flares in the bad pile, what would your strategy be if you needed seven good flares? What would your probability of success be?*

Do no testing. Pick a pile at random. You have a 1/2 chance of choosing the good pile. If you do, then you have a further 1/2 chance of choosing a good flare from the pile having three bad flares. These are independent, so the probability you collect seven good flares is 1/2 × 1/2 = 1/4.

4. *If there were four bad flares in the bad pile but you needed seven good flares, then what would your strategy be?*

You follow the same strategy, but now the probability of success is 1/2 × 2/6 = 1/6. So having four bad flares reduces the probability of success.

5. *Is there anything strange going on here?*

It's a bit strange that increasing the number of bad flares increases your probability of success when you seek only five flares but decreases your probability of success when you seek seven.

6. *In the case where the bad pile has three bad flares, you could still take all 12 and get five good ones. You could even take five from each pile and be sure to get five good ones. Can you design a method that will enable you to take no more than seven and have five good flares no matter how the flares are distributed? How about a strategy to enable you to take no more than eight?*

You might have to take as many as eight. Here's why. Suppose that whenever you test, you get a good flare unless you are testing the bad pile and there are three or fewer left. So, let's say that you test four in pile A. If none are bad, then the other two in pile A are good, but then you need all six from pile B. If you test any fewer, under our supposition, you have learned nothing. So, suppose you test none. Then you must take all of pile A and two from pile B (or vice versa). This is guaranteed to work because either all six from A are good or three from A are good and the two from B are good.

7. *Suppose that you insist on guaranteeing that you never have to carry more than the number you found in your answer to the previous question and further guaranteeing that you have five good flares. How can you nevertheless guarantee that most of the time you need only carry five with you?*

If you test one from each pile then you have a 50 percent chance of finding the bad pile, so you can take all the others from the other pile. Test one from pile A and one from pile B. If either is bad, you take five from the other pile and you are done. The likelihood of this is $(1/2 \times 1/2) + (1/2 \times 1/2) = 1/2$. Otherwise, test a second one from pile A. If it is bad, then take all five from pile B (giving you an additional probability of $(1/2 \times 1/2 \times 3/5)$ so you get a total of 13/20). If good, then take four from each pile. No matter what happens, you will get five good flares.

THE COLOR FAIRIES **SOLUTION**

Which fairy is attracted to which color?

A. Neither Anya, Caroline, Oliviana, or Ariana can like rose, turquoise, or violet. Reason: Tyler receives nothing on nights 2 and 3 when those are the active fairies.

B. Cloe likes either rose or turquoise. Reason: Tyler, night 1 compared with night 2 and A.

C. Oliviana likes either ivory or sage. Reason: Jordan, night 1 compared with night 2 and A.

D. From A and the fact that David gets two in both nights 1 and 2 we know that David gets two for the colors sage and emerald. From B, we know that Cloe doesn't like sage or emerald. So Anya and Caroline like sage and emerald but we don't know who likes which.

E. From night 4, when David gets none, we know that Cloe, Ariana, and Oliviana dislike sage, violet, and emerald.

F. From night 5, when David gets one, we know that Anya likes either sage or emerald.

G. From Jordan and night 2 compared with night 3, we know that Oliviana likes either ivory or sage (not violet because of A). However, she cannot like sage because either Anya or Caroline likes sage. Therefore, Oliviana likes ivory.

H. From Jordan, night 3 compared with night 4, and G, we know that Oliviana likes ivory and Anya must play a role. Since Anya likes either sage or emerald from D, Anya must like sage.

I. From H and D, Caroline likes emerald.

J. Ariana likes either silver or gold or earth, because she doesn't affect the pearl counts in nights 3 or 4 by G, H, and I.

K. Cloe likes either rose or turquoise.

L. Since someone likes turquoise and someone likes earth from point 3 of "what you know," we conclude that Ariana likes earth and Cloe likes turquoise.

ALTERNATING LIARS **SOLUTION**

1. You have five people to choose among, you are allowed to ask only two questions (of the same or different respondents). One must be a yes/no question, but the other may ask the respondent to point to at most one person. Can you find the consistent truth teller?

Ask A, "Are you the consistent truth teller?" If A answers yes, then ask A, "Who is the truth teller?" If A points to A, then A is it. If A points to anyone else, then that person is the consistent truth teller because A has given inconsistent answers, so cannot himself be the consistent truth teller. Therefore, A must have lied the first time and therefore is telling the truth when pointing elsewhere.

If A answers no, then A is telling the truth but is an alternating liar so ask "Please point to one person who is an alternating liar." The result will be a lie and therefore will point to the person you seek. This solution in fact works for any number of people.

One subtle point should be noted: It does not work to ask person A in the room "Who is the truth teller?" "Who is the truth teller?" because A could answer B to the first and C to the second. Either could be a lie.

2. *Suppose you were allowed to broadcast your questions to all five people and get an answer from each one. But the questions have to be yes/no. Could you do it with two broadcast questions?*

Ask everyone twice: "Are you a truth teller?" Only those who respond yes both times are real truth tellers.

This works for any number of people. If you are limited to two yes/no questions but without broadcast (that is, each question must be addressed to and answered by one person), then are two questions possible? I don't think so, but I don't know for sure.

3. *How many questions would you need if there were seven people in the room?*

Both of the above solutions work for seven people.

4. *Suppose there were only two people in the room, one a consistent truth teller and one an alternating liar. Could you determine who was who with one yes/no question to one respondent?*

Ask "Is exactly one of the following true: (*i*) you are the consistent truth teller, (*ii*) you are in a lying phase?" The truth teller will say yes. An alternating liar who is telling the truth will say no. An alternating liar who is lying will notice that exactly one is true, but the alternating liar lies, so he will say no.

NOTE TO THE ADVANCED READER.

When this puzzle was first published, the question "Can one pointing request identify the consistent truth teller in a room with five people?" sparked a vigorous debate. I think that discussion bears repeating, because the last word has not yet been spoken, I think.

The most intriguing answers were proposed by Joao Borges de Assuncao, Bruce Tattrie, M. E. Fisher, Dan Garcia, and Jon Hamilton. The Garcia formulation goes like this: " 'If the next question were to choose someone who was the same (truthteller or alternating liar) as you, point to one person you *could* choose.' The truthteller will point to him/herself. The truth-first alternating liar will say 'Ok, next time I'll lie. As a liar, if I were asked to point to someone the same as me, I'd have to choose the truthteller. But since I'm now telling the truth, I have to report truthfully that I would point to the truthteller.' These folks point to the truthteller. The lie-first alternating liar will say 'If, next time as a truth-telling alternating liar I were asked to choose someone the same as me, I could choose any of the other alternating liars. But I have to lie now. So I have to choose someone I could NOT pick next time, which must be the truthteller.' These folks point to the truthteller."

Mark-Sibley Schreiber points out that since there is no next question, the precondition part—"If the next question were to choose someone who was the same (truthteller or alternating liar) as you, . . ." is not true. Therefore, the respondent should be free to point to anyone at all. But maybe the prisoners haven't been trained in such refined logic.

ALPHA MALOS **SOLUTION**

Given that

C speaks to F,

B speaks to E,

A speaks to F,

F speaks to A,

E speaks to B,

A speaks to B,

F kicks D,

E kicks A,

A kicks C,

A kicks B,

B kicks A,

A kicks E,

B kicks C,

C kicks A,

who sported with whom?

C sported with A's woman,

D sported with F's woman,

C told F that D sported with F's woman,

B told E that A sported with E's woman,

A told F that D sported with F's woman,

F told A that C sported with A's woman,

E told B that A sported with B's woman,

A told B that C sported with B's woman.

EXPLANATION.

- Since D kicks nobody, D must have sported with F's woman. This accounts for at least one of the speaking events: "C speaks to F," "A speaks to F." Since F kicks nobody else, it must be both.

- E kicks A, but A may be innocent of sporting with E's wife since he kicks B, C, and E, and B spoke to E. So, perhaps A is innocent and B has accused falsely.

- A kicks C and C kicks only A. Therefore, C must have sported with A's woman and the accusation came from F.

- A kicks B because B was a false accuser.

- B must kick A for one reason only: E accused A of sporting. Since A later kicks E, A is claiming that this was a false accusation.

- The last two kicks are due to A's false claim that C sported with B's woman.

NOTE TO THE READER.

After healthy debate, *Scientific American* decided this was too much for a magazine audience. Please take it as it is intended—fun and satire.

DISPOSABLE COURIERS **SOLUTION**

1. There is a kindred warm-up problem in which the secret is a line. Which information would you then give each of your three friends?

Each person gets a different point on the line. Any two points define the line, but one tells you nothing about the line.

2. Now try either approach for the five-courier problem.

There are two approaches to the main problem. First, in the spirit of the solution to the warm-up: encode your message as a number that will then become, say, the x-coordinate of a point P in three-dimensional space. Choose two other coordinates randomly. Now select five planes that all intersect at point P and assign each courier to a different plane, defining it for the courier using the coordinates of three other (non-P) points in the plane.

Two nonparallel planes meet at a line (as shown in Figure S1), and any plane not containing that line but intersecting it will hit the line at a single point. So, knowing the planes of any two couriers

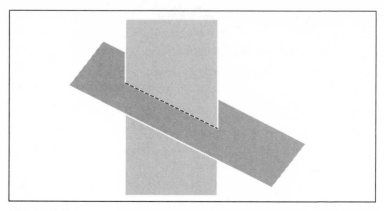

Figure S1 The intersection of two non-parallel planes is a line as shown. The intersection of that line with a third plane that is neither parallel to it nor contains it, is a point.

would give the enemy no information about point P except the line on which it lies, but any three couriers together could find the critical point readily and determine the x-coordinate to uncover the full message.

The number representing your message—the x-coordinate—will probably be very big, but for the sake of explanation, let's say it is just 317. And let's say the other coordinates you choose are 211 and 894, yielding the point $P = (317, 211, 894)$. You can construct the five planes by defining 10 lines containing point P such that no three lie in the same plane. Any pair of the lines will define a plane (as shown in the Figure S2), giving you five planes from the 10 lines. If you define those planes using non-collinear points that do not include $x = 317$, any two couriers who combined their planes would meet at a common line that would include $x = 317$, but they would have no way of knowing that 317 was any more informative than any other of the infinite points on the line. On the other hand, any third courier would intersect the other planes at point P, revealing the x-value.

Clever readers will note a deeper issue with this problem. Whereas the specific point is hidden in a geometrical sense until

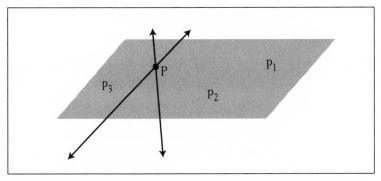

Figure S2 Suppose that the dark circle represents the point P, whose x coordinate encodes the message. The two lines going through P define the plane shown. The courier corresponding to this plane could be given three points in this plane that are off either line, shown here as P_1, P_2, and P_3. Capturing a single courier would give information about the entire plane, an infinite number of points. Altogether, we could create five such pairs of lines, all going through point P, no three of which are co-planar.

three couriers appear, the message may not be. Given a line or even a plane, one can try many integer x-coordinates until a plausible rendezvous time and location appear. Knowing inside information about the message (for example, it's in English) can help disambiguate a message.

The original reference to secret sharing is Adi Shamir's "How to Share a Secret," *Communications of the ACM* 22, no. 11 (November 1979): 612–613, which formulates the problem as one about finding the zeros of polynomials. It uses the fact that for a polynomial of degree d, if we have d or fewer points, we cannot find the zero. For our problem, that brings us to the second manner of solving the problem, which is probably better. Three points are enough to describe a parabola, so we give each courier a different point of the parabola. Any two tell very little, but the three together completely characterize the curve. This is analogous to our solution of problem 1.

NOTES.

Disappearing Cryptography by Peter Wayner (published in 2002 by Morgan Kaufmann Publishers) is a friendly introduction to this field. A more advanced reference is Bruce Schneier's *Applied Cryptography* (John Wiley and Sons, 1996). Carl Bosley of New York University and David Molnar at Berkeley contributed several insights to this discussion.

THE DELPHI FLIP **SOLUTION**

1. How do you end up with the greatest possible final amount, no matter when the oracle chooses to lie? How would that change if you had 20 bets?

When you decide on the wager amount as you go, you will end up with slightly more than $17,000. Here is why. Let $S(x,t,n)$ be defined as the amount of money you will have at the end of the betting session if you have x money now, you have t turns left, and the malicious oracle has n more lies to make (n is either 1 or 0).

Suppose that at some point you have t turns to go, the oracle has no more bad things to do, and you have x dollars. Then you bet everything each time: $S(x,t,0) = (2^t)x$

Suppose that you have one turn to go and the oracle has one bad thing to do. Then you bet nothing: $S(x,1,1) = x$.

Now, let us try to solve for $S(x,t + 1,1)$. If we bet b, then either the oracle gets us—in which case we will end up with $S(x - b,t,0)$— or we will have $x + b$ and we will end up with $S(x + b,t,1)$. We want $S(x - b,t,0) = S(x + b,t,1)$ otherwise the malicious oracle will make us choose the worst outcome. So, we use what computer scientists call dynamic programming:

$S(x,1,1) = x,$
$S(x,1,0) = 2x,$
$S(x,t,0) = 2^t x,$
$S(x,2,1) = $ bet $x/3$ and end up with $4x/3,$

$S(x,3,1) =$ bet $x/2$ and end up with $2x$,

$S(x,4,1) = 2(x + b) = 2x + 2b = 8(x - b) = 8x - 8b$, where b is the bet,
  so $10b = 6x$, so $b = (6/10)x$.
  The result is $(32/10)x$.

$S(x,5,1) = 3.2(x + b) = 16 - 16b$.

And so on. It's best to do this with a computer program. Here are the sums you will end up with after each flip:

After (flips)	Amount You Have (At Least)
2	$133.3333
3	$200
4	$320
5	$533.3333
6	$914.2857
7	$1,600
8	$2,844.444
9	$5,120
10	$9,309.091
11	$17,066.67

After 20 flips, you'd have nearly $10 million.

2. *Suppose you have to decide the amount of all of your bets in advance without knowing when the oracle will lie. What should your bets be in that case and what final amount can you be sure to get no matter when (and if) the oracle chooses to lie? (You lose everything if you plan for a bet on a particular move but end up having too little money at that time.)*

Now, if you must bet in advance, you will end up with only $1,600.

Here is a betting strategy that achieves that:

$50, 50, 100, 100, 200, 200, 400, 400, 800, and 800.

That is, $50 the first and second rounds, then $100 the third and fourth rounds, etc. This gave a guaranteed final result of $1,600 as follows (these are the lowest results after each bet assuming there has been a lie):

$50; 100; 100; 200; 200; 400; 400; 800; 800; 1,600

You'll note that at every time, you never bet more than half of what you have. We will come back to this theme in later puzzles.

CROWNING THE MINOTAUR SOLUTION

1. What is your strategy if there are three prisoners?

Each prisoner follows the rule: If I see two reds, I bet blue. If I see two blues, I say red. Otherwise I pass. Clearly somebody must not pass because there will always be at least two reds or two blues. Suppose there are at least two reds. Then there are four possibilities. A and B alone are red: A and B pass. C will say blue. Correct. B and C alone are red: B and C pass. A will say blue. Correct. A and C alone are red: A and C pass. B will say blue. Correct. A, B, and C are red: All will say blue and all will be incorrect. So, the prisoners will win ¾ of the time when there are at least two reds and at least ¾ of the time when there are at least two blues.

2. What is your strategy when prisoners can bet a different number of points?

The prisoners agree on a protocol based on who stands where. The prisoner at Athena always bets one point for red. If the prisoner at Poseidon sees that the Athena prisoner has a blue crown, then the Poseidon prisoner bets red but with two points and otherwise passes. If the Zeus prisoner sees that the first two prisoners both have blue crowns, then the Zeus prisoner bets red but with four points and otherwise passes. Unless all three prisoners have blue crowns (a 1 in 8 chance), the prisoners will win.

3. How do the odds change as the number of prisoners increases?

The odds get better as the number of prisoners increases, but the math gets significantly more difficult. For seven prisoners, here is a strategy that ensures 7/8 odds of winning. To represent the problem, denote red by 0 and blue by 1. Now consider the "binary Hamming 7,4,3" code:

0	0000000
1	0001110
2	0010101
3	0011011
4	0100011
5	0101101
6	0110110
7	0111000
8	1000111
9	1001001
10	1010010
11	1011100
12	1100100
13	1101010
14	1110001
15	1111111

These bit representations have the following properties:

i. Every two of these "code points" differ from one another in at least three places.

ii. Every possible seven-bit number is either a "code point" or differs from a code point by one bit. The neighborhood of each code point will consist of that number, as well as all numbers that differ from the code by one bit.

For example, the neighborhood of 0110110 is

0110110
1110110
0010110
0100110
0111110
0110010
0110100
0110111

iii. No neighborhoods from different code points overlap.

Now, a random seven-bit number is most likely (probability $112/128 = 7/8$) not to hit a code point. The seven prisoners conspire to win by this probability as follows.

Say the crowns are distributed somehow, for example, 0010110. Because of the above facts, this number is either a code point or is one bit away from a code point. As it happens, it is one away from 0110110 in that it differs by the second bit. Of course the second youth (the one next to Aries) doesn't see that bit position which corresponds to his crown. Instead, he sees 0?10110. So, he thinks the configuration can be either 0110110 (code point) or 0010110 (not a code point). He will guess—correctly in this case—that it is not a code point by deciding that his crown color is zero (red). In this case, only youth 2 will make that guess.

On the other hand, if the configuration is in fact the code point 0110110, then youth 1 will see ?110110 and will guess blue (incorrectly). Youth 2 will see 0?10110 and will guess red (incorrectly). Youth 3 will see 01?0110 and will also guess red also incorrectly, and so on. So, whenever the configuration is a code point, the players will lose.

4. For the seven-prisoner case, how does the strategy change if prisoners can bet points?

If each player can choose the number of points to bet on his guess and if the majority of points cast are correct, then the players win and otherwise they lose. They then follow these rules: Suppose the first player (next to Apollo) is player 1. Player 1 guesses red. Player k will pass unless all $k - 1$ previous players have blues. In that case, player k will bet 2^k points for red. Only if all players get blues will they lose. In this case, that is 1 chance in 128.

MATHEMATICAL NOTE.

The essential structure of this puzzle (except for the majority points part which may be new) came from Todd Ebert's Ph.D. thesis in 1998 and was presented by Peter Winkler at the 2000 Gathering for Gardner conference. It should also be noted that in 1987 Steven Rudich and Richard Beigel posed and published the solution to a similar problem related to voting. Yevgeniy Dodis explained this solution to me and referred me to the concept of the binary Hamming code. Andy McDaniel (and his friends Sharon and Tom) observed an ambiguity about ordering in the original formulation which is fixed here by the use of statues.

SAFETY IN DIFFICULTY **SOLUTION**

What information should the guards receive and how should the spies present their passwords so that only your spies get through—even if the guards go out for a couple of pints?

We want a one-way function such that the guard's verification job is easy, but the adversary's job (finding the inverse function) is very hard.

A prime-based strategy would work as follows: Before leaving the home country, the spy generates two large random prime numbers p and q and gives the border guards the product $p \times q$. The guards do not need to know who told them this, just that some legitimate spy did. When a guard challenges the spy upon return, the spy gives the guard the two prime numbers; the guard multiplies them and sees that the product matches the product on his list. Of course these numbers cannot be used again by this or any other spy, so the guard tells all other guards to eliminate this number from the lists. Even if a guard reveals one of the unused products in a bar, it is infeasible to find the prime factors, so infiltration by an enemy spy is still not possible.

Here is the solution Rabin suggested. He used a function first introduced by John Von Neumann for the purpose of generating pseudo-random numbers: Start with some starting number (the first number of the sequence) and square it. Then take the middle digits of the result as the second number of the sequence and square that. Continue in that fashion to generate a pseudo-random sequence.

Rabin's method worked as follows. Have each spy remember a large randomly chosen number N of say 50 digits. When challenged, the spy will report N. The guard computes $M = N \times N$ then takes the middle 50 digits of M. The guard checks that the middle 50 digits are on his list and not crossed out. If so, the guard lets the spy pass and crosses out that number and tells the other guard stations. Next time spy s goes on a mission, he generates another random number and the middle 50 digits are sent to the guard stations.

Even if the guard gives up his entire list, the adversary must do the equivalent of finding N given the middle 50 digits of M. That is extremely hard, though the mathematically adept reader will note that several N's may work.

MATHEMATICAL NOTE.

To learn more about progress in factorization, go to www.loria.fr/ ~zimmerma/records/gnfs158.

FISHY BUSINESS **SOLUTION**

1. Can you find the one bungalow in which the fish vandal must be hiding?

Only D and G are touched by an odd number of pathways. If one is the starting bungalow, then the other must be the end bungalow, because any bungalow touched by an even number of pathways will have as many trips out as in. If neither D nor G is the starting bungalow, then consider the one that is not the ending bungalow either. Call it X. The fish vandal will enter X by a path and leave X by a different path each time. Because there are an odd number of paths, however, he must leave one path empty. This contradicts the evidence. Hence D or G is the starting bungalow. Because only D is on the lagoon side, the vandal must be in G (see Figure S3).

2. What if the fish vandal could retrace at most one path but must still cover all paths and still comes in on the lagoon side. Then where might he be?

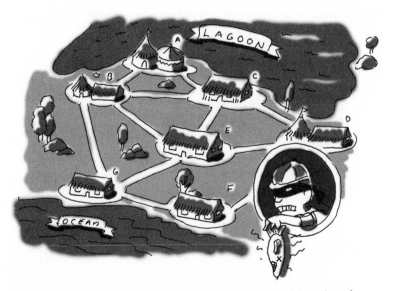

Figure S3 Observe that only D and G are touched by an odd number of pathways.

Tomas Rokicki has suggested the conceptualization that retracing has the same effect as cloning a path between two existing bungalows. If one clones a path between two bungalows having an even number of paths (say B and E), then four bungalows would have an odd number of paths touching them. Because a bungalow with an odd number of paths must be the beginning or end of the fishmonger's tour, this is not possible. The two bungalows that are already touched by an odd number of paths (D and G) have no direct path between them. So the only other possibility is a path starting from a bungalow touched by an odd number of paths to another bungalow touched by an even number of paths. Those are GB, GE, GF, DC, DE, and DF.

The starting point must be one of B, A, C, or D because the fishmonger comes from the lagoon. If it is D and the fishmonger retraces a path from G at the very end, the fishmonger can end up at B, E, or F. Surprisingly he can also end up at D. Here is how: The extra path is GB. He starts at B, goes to G, comes back to B. At that point B has three unused paths touching it and G has two. So the only possible endpoint of this path will be D.

MATHEMATICAL NOTE.

The fundamental mathematical theorem here is known as the Euler Trail theorem. An Euler Trail is a walk through a graph that uses each edge exactly once. The theorem is that every graph with two or fewer odd-degree vertices (for us, a bungalow touched by an odd number of paths) has at least one Euler Trail.

PERFECT BILLIARDS **SOLUTION**

1. Can you make the shot by banking the ball twice against the cushions? At what slope should you hit the ball?

Yes, hit the ball at a slope of ½ in an east-northeast direction. The best way to see this is to imagine three rectangles: the original one, another one immediately to the right, and one above the second one. Now imagine a line from (2,0) on the first rectangle to (3,1) on the upper-right rectangle. Such a shot must travel four meters to the right and two meters up. That can be done only with a slope of ½. Now observe that the reflections will bring that same ball to the near corner (see Figure S4).

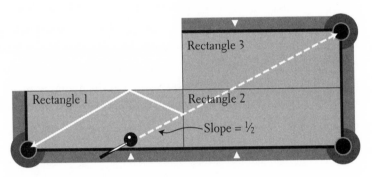

Figure S4 In going from (2,0) in rectangle 1 to (3,1) in rectangle 3, we are leaving at the same angle as the ball would need to travel to end at (0,0). That is a slope of ½.

However, other solutions are possible. Richard Pineo and Roy Freborg proposed the following out-of-the-box solution. Recall that the ball is at position (2,0), so on the south side. Nevertheless, one can imagine that it's one micron off the side or that the cushioning on the side bends a little. In that case, one can build on the one-bank solution as follows: Hit the ball southwest at a slope of 1. The ball will bank against positions (2,0) and (1,1) before gliding into the pocket.

Some players advocate rail shots that would bounce twice in the target pocket, but I couldn't confirm whether that could be guaranteed and at which precise angle.

2. For those who are particularly strong in geometry, consider what would happen if the ball were initially at some arbitrary point on the southern edge. How would you discover the angle to shoot using only the ability to draw line segments and parallel line segments and to discover the midpoint of a line segment?

Suppose we have any rectangular table and the initial ball can be anywhere on the southern side. Figure S5 shows the distorted geometrical drawing that goes with the argument of clever reader David Wilkinson: Imagine a rectangular billiard table *ABCD* with arbitrary aspect ratio and pockets only in the corners. There is just one ball located at an arbitrary point *P* on side *AB*. The goal is to hit the ball off the (adjacent) side *BC*, then off the (opposite) side *DC* and into the pocket at *A*. Now, choose a point *S* which is the midpoint of *AP*, and hit the ball parallel to *SC* at an angle we might call *X*. Draw a line from *P*, parallel to *SC*, meeting *BC* at *Q*, and *DC* at *T*. Also, draw a line from *A* parallel to *SC*, meeting *DC* at *R*.

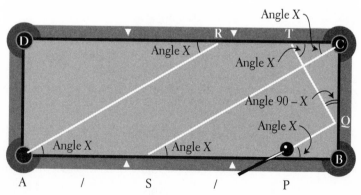

Figure S5 Though the drawing shows *T* and *R* to be distinct, we will prove them to be identical.

ASSERTION.

$PQRA$ is the path of the ball.

PROOF.

Clearly $AS = RC$, because these are opposite sides of a parallelogram. Now, we will also show that $SP = CT$, but this takes some more work (unless you remember your theorems in geometry really well). We notice that $\angle SCB$ is similar to $\angle PQB$ and to $\angle TQC$. Therefore, $CT/CQ = SB/BC = PB/BQ$. Now, $CT = SB \times CQ/BC$. $SP = SB - PB = SB - (SB \times BQ)/BC = SB(1 - BQ/BC) = SB((BC - BQ)/BC) = SB \times CQ/BC$. Hence $CT = SP$. But $AS = SP$ by construction, so $RC = CT$ (from the first sentence of the proof) and hence R and T are the same point. Further, as the diagram shows, the angle QTC is exactly X, so the ball from P to Q to T ($= R$) will reflect into A.

GHOST CHOREOGRAPHY SOLUTION

1. Can you find a two-step swap for any even number of dancers?

    00000000
    xxxxxxxx

can be transformed to

    xxxxxxxx
    00000000

in two steps satisfying the swapping prohibition for any even number. Consider each foursome:

    xx
    00

    xo
    xo

    00
    xx

2. *Suppose we start with*

xxxxxxxx
xxxxxxxx
oooooooo
oooooooo

and want to get

oooooooo
oooooooo
xxxxxxxx
xxxxxxxx

How do we do that?

Take each eightsome

oo
oo
xx
xx

and proceed as follows:

oo
xo
xo
xx

xo
xo
xo
xo

xx
xo
xo
oo

and finally

```
xx
xx
oo
oo
```

Surprisingly, the end column can do better:

```
x
x
o
o
```

```
 x
ox
o
```

```
o
ox
 x
```

```
 o
 o
 x
 x
```

which suggests the following better setup for pairs:

```
xx (1)
xx (2)
oo (3)
oo (4)
```

```
  xx  (2)
oxxo (3)
o  o (4)
```

```
o   o (2)
oxxo (3)
 xx  (4)
```

```
oo (2)
oo (3)
xx (4)
xx (5)
```

3. *Can you design the moves of the dancers at each step to go from x's surrounding o's to o's surrounding x's?*

```
   xx    (1)
  xxxx   (2)
 oooooo  (3)
  oooo   (4)
  oooo   (5)
 oooooo  (6)
  xxxx   (7)
   xx    (8)
```

```
 x   x   (1)
oxooxo   (2)
oxooxo   (3)
o     o  (4)
o     o  (5)
oxooxo   (6)
oxooxo   (7)
 x   x   (8)
```

```
     oo     (1)
   oxooxo   (2)
   ox  xo   (3)
   ox  xo   (4)
   ox  xo   (5)
   ox  xo   (6)
   oxooxo   (7)
     oo     (8)

     oo     (0)
     oo     (1)
    oxxo    (2)
    oxxo    (3)
    oxxo    (4)
    oxxo    (5)
    oxxo    (6)
    oxxo    (7)
     oo     (8)
     oo     (9)
```

TRUCK STOP **SOLUTION**

1. If two lanes can be blocked, how fast can the delivery trucks reach their destinations?

Here is a six-minute solution: Outer lanes (notation AE(A,B) means the vehicle going from A to E went in this time period from A to B). The symbol _done means that the trip is finished. For example, AD(C,D)_done in time 4 below means that the truck from A to D finally reached D, passing through (C, D).

Outer lanes:

Minute 1: AE(A,B), BE(B,C), CE(C,D), DE(D,E)_done;

Minute 2: AE(B,C), BE(C,D), CE(D,E)_done, AD(A,B);

Minute 3: AE(C,D), BE(D,E)_done, AD(B,C), AC(A,B);

Minute 4: AB(A,B)_done, AE(D,E)_done, AD(C,D)_done, AC(B,C)_done;

Minute 5: BD(B,C), CD(C,D)_done;

Minute 6: BD(C,D)_done, BC(B,C)_done.

Inner lanes:

Minute 1: EA(E,D), DA(D,C), CA(C,B), BA(B,A)_done;

Minute 2: EA(D,C), EB(E,D), CA(B,A)_done, DA(C,B);

Minute 3: EA(C,B), EB(D,C), EC(E,D), DA(B,A)_done;

Minute 4: EA(B,A)_done, EB(C,B)_done, EC(D,C)_done, ED(E,D)_done;

Minute 5: DB(D,C), CB(C,B)_done;

Minute 6: DB(C,B)_done, DC(D,C)_done.

2. *Can you prove that yours is one of the fastest possible solutions?*

This is in fact a minimum: consider the lane BC. How many must pass it? AE, AD, AC, BD, BC, BE, EA, EB, DA, DB, CA, CB. So, 12 trips traverse BC. Therefore there is no way to do better than six minutes.

3. *Could the strikers force the deliveries to take more time by blocking two other lanes instead?*

By symmetry, blocking any two lanes along any single side (for example, blocking the two lanes between C and D) yields the same situation as in the second solution. So, the two questions are whether (*i*) blocking two lanes on neighboring sides or (*ii*) blocking two lanes on non-neighboring sides would delay things more. Luckily we can treat the two cases in the same way for the purposes of this question.

Suppose we blocked the inner lanes between A and B and between B and C. The most isolated element is B. Use the outer lanes only as in the solution to the warm-up so everyone sends to right neighbors both one and two away. This takes three minutes. Then use the outer lanes to follow the inner-lane solution to the warm-up. So everyone sends to left neighbors both one and two away. This takes another three minutes. So, six minutes is doable.

In fact, they could block the entire set of inner lanes and all goods could still be delivered in six minutes. We know this because we didn't use the inner lanes. So, clearly, the strikers can't slow things down more by blocking two lanes that aren't on the same side than by blocking two lanes on the same side.

4. Would any two-lane blockage demand less time than six minutes?

Suppose the strikers, in their attempt to show strength without antagonizing people too much, block two lanes on adjacent sides. Say they blocked the inner lanes between A and B and between B and C. The most isolated element is B. So, B sends right by one and then by two and, in parallel, left by one and then by two. After three minutes, B's sending obligations are settled. Also, during that time E has sent to A the vehicle intended for B. D has sent to C the vehicle intended for B. So in the two minutes starting at minute 3, C sends to B the two vehicles from C and D. Symmetrically, A sends to B the two vehicles from A and E. So, B could be done after four minutes. We have to dispatch all the other obligations of all other sites now.

We summarize all this as follows. (Recall that DB(D,C), for example, means that the truck from D to B takes the lane to C. Similarly, BE(A,E)_done means that the truck from B to E takes the lane from A to E.)

Minute 1: BA(B,A)_done, BC(B,C)_done, EB(E,A), DB(D,C);

Minute 2: BE(B,A), BD(B,C);

Minute 3: BE(A,E)_done, BD(C,D)_done, AB(A,B)_done, CB(C,B)_done;

Minute 4: EB(A,B)_done, DB(C,B)_done.

Let's see if we can take care of everyone else by using the unused lanes. In minute 1, while B is sending left and right, A sends to E its vehicle destined for C: AC(A,E). Symmetrically, C sends to D its vehicle for A: CA(C,D). At the same time D sends to E its vehicle for A and E sends to D its vehicle destined for C: DA(D,E) and EC(E,D). So, minute 1 is extended to use all possible lanes.

Minute 1: BA(B,A)_done, BC(B,C)_done, EB(E,A), DB(D,C), AC(A,E), CA(C,D), DA(D,E), EC(E,D).

In the next minute, we continue the trips from A to C and from C to A and begin working on some two-away trips from A to D and from C to E. We also finish the two-away trips from E and from D.

Minute 2: BE(B,A), BD(B,C), AC(E,D), CA(D,E), CE(C,D), EC(D,C)_done, AD(A,E), DA(E,A)_done.

In minute 3, all two-away and three-away trips from A and C finish so only one-away trips remain.

Minute 3: BE(A,E)_done, BD(C,D)_done, AB(A,B)_done, CB(C,B)_done, AC(D,C)_done, AD(E,D)_done, CA(E,A)_done, CE(D,E)_done.

We do the one-away exchanges in the last minute while doing final deliveries to B.

Minute 4: EB(A,B)_done, DB(C,B)_done, EA(E,A)_done, AE(A,E)_done, DE(D,E)_done, ED(E,D)_done, CD(C,D)_done, DC(D,C)_done.

There is no way to do better because B alone has to send to four other sites and receive from four other sites and so requires four minutes.

TRUST NO ONE **SOLUTION**

1. What is the fewest number of tests you can use to verify the four-element circuit, and which test or tests would that entail?

You can use as few as two tests to verify the four-element circuit:

   i. $A = B = C = 0$ and $D = 1$. Output should be 1.

  ii. $A = C = 1$ and $B = D = 0$. Output should be 0.

The first situation could come about from the following circuit element assignments:

elem0	elem1	elem2	elem3
AND	AND	OR	OR
AND	OR	OR	OR
OR	AND	OR	OR
OR	OR	OR	OR

The second situation could come about from:

elem0	elem1	elem2	elem3
AND	AND	AND	AND
AND	AND	AND	OR
AND	AND	OR	AND
AND	AND	OR	OR
AND	OR	AND	AND
OR	AND	AND	AND
OR	AND	OR	AND
OR	OR	AND	AND

Only the desired circuit would satisfy both.

2. What if the desired circuit had four ANDs?

In that case, three tests are required: 0, 1, 1, 1 (meaning $A = 0$ and the rest are 1); 1, 0, 1, 1; and 1, 1, 1, 0.

The first test could result from these circuits:

elem0	elem1	elem2	elem3
AND	AND	AND	AND
AND	AND	OR	AND
AND	OR	AND	AND
AND	OR	OR	AND

The second from these:

elem0	elem1	elem2	elem3
AND	AND	AND	AND
AND	AND	AND	OR
AND	AND	OR	AND
OR	AND	AND	AND

The third from these:

elem0	elem1	elem2	elem3
AND	AND	AND	AND
AND	OR	AND	AND
OR	AND	AND	AND
OR	OR	AND	AND

3. *Using the interconnection of the warm-up, which combinations of box (logic element) values can be verified by using just one test?*

Any circuit with the interconnection of the warm-up can be verified with one test except those in which elem 1 is an AND when elem 0 and elem 2 are ORs or in which elem 1 is an OR when elem 0 and elem 2 are ANDs. To see this, suppose both elem 0 and elem 2 should be OR. Then any test for them cannot set both A and D to be 1, because then the output of elem 1 doesn't matter. Suppose then that A is 0 and D is 1. The intended output from elem 0 is 0 and from elem 1 is 1. Suppose that one of B and C is 1 (if they are both 1, then elem 2 could be an AND or an OR and it would still yield an output of 1). Then the output of 0 from elem 0 could be due either to the fact that elem 0 is an AND or that elem 1 is an AND. See if you can work this out from here.

PEBBLING A POLAR BEAR **SOLUTION**

1. Are four gunners enough? If so, show how.

Four gunners are not enough. Consider the slight modification of the wheel pattern (see Figure S6). The center must always be guarded, otherwise the bear can run through the center to an unguarded other node. On the other hand, five are enough for this pattern, because the gunners can enter in the center, leave one gun-

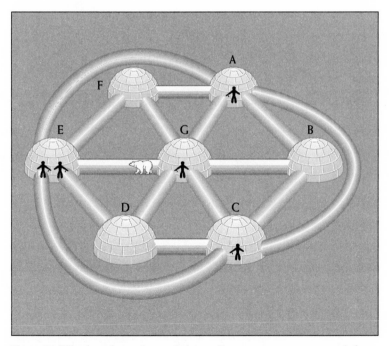

Figure S6 Wheel, spoke, and extra linkages. Four gunners are not enough for that.

ner there, and move to any highly connected perimeter node, say A. Leave one guard at A. Move three to C. Leave one guard at C and move with the remaining two to E.

2. If not, show how many you need and that it will work.

In fact, five are enough for any seven-node planar graph because every such graph has at least one node N with four or fewer neighbors. Case 1: N has four neighbors (see Figure S7). One can send gunners to all four neighbors of N, guard three of them, enter the fourth neighbor N¹ with a party of two and then send those to N and back to N¹. The net effect will be that N will be unguarded but will be guaranteed free of bears. The four neighbors of N will be guarded and hence free of bears.

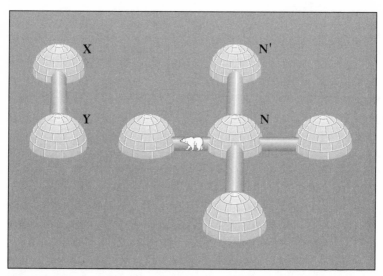

Figure S7 There must be some node N having four or fewer neighbors. The remaining two nodes (X and Y) cannot form triangles with all four neighbors of N. So send the search party from the node to which they don't form triangles and the polar bear will be found.

Now consider the remaining two nodes X and Y. Say that X and Y have a triangle with respect to Z if there are corridors between X and Y, Y and Z, and X and Z. By inspection, there must be one neighbor M of N such that X and Y have no triangle with respect to M. Start from M with a party of two and search for the bear in X and Y.

If there is no node with exactly four neighbors but one node T with exactly three neighbors, go to the neighbors of T, guard them, then visit T. At that point, T is unguarded but bear-free. Let the neighbors of T be A, B, and C. Let the remaining igloos be X, Y, and Z. It is a fundamental fact about planar graphs that at least one of X, Y, Z has no corridor to one of A, B, C. Say that X has no corridor to C. So, leave a guard at C and move to either Y or Z or possibly both with two gunners. At that point C can be left unguarded and a pair of guards can move to X.

3. *Suppose there were 100 igloos arranged in a rectangular grid, where each igloo has corridors to its (at most four) grid neighbors, horizontal and vertical. What is the smallest number of gunners that is guaranteed to be enough?*

The worst case is that the grid is a square in which case 11 gunners (one more than the square root of 100) are enough. In a rectangle of the form 5×20, say, then six gunners would be enough. In general if the rectangle is $m \times n$ where $m \leq n$, then occupy the m positions on the far left. Advance the bottom occupant by one to the right, then the next to the bottom, and so on. One needs $m + 1$ gunners (see Figure S8).

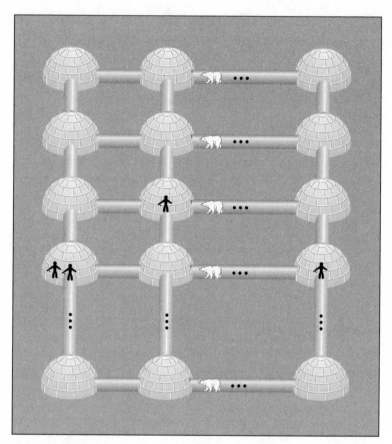

Figure S8 If the igloos form a large matrix, even with fixed degree, one needs a number of gunners proportional to the smaller of the number of rows or the number of columns to stun the bear.

THE GRAPH OF LIFE **SOLUTION**

1. Given only two groups, two subscripts per group, where each subscript can be either 1 or 2, can you add any species to the example of the last part of the warm-up and still get derivation by a tree? For example, could you add some different species of the form A_ij B_km?

No, because any more would create "conflicting pairs." For example, suppose you added

A_22 B_21.

To get the five species

S1: A_12 B_21,
S2: A_11 B_22,
S3: A_22 B_12,
S4: A_21 B_11,
S5: A_22 B_21,

would imply that we would have to account for two pairs of species:

Pair 1
S5: A_22 B_21,
S3: A_22 B_12,

and

Pair 2
S5: A_22 B_21,
S1: A_12 B_21.

Pair 1 requires that some species in the tree is A_22 B as the arise point of A_22. Specifically, B cannot differentiate into variants B_1 and B_2 before A_22 arises. The reason has to do with rule 3. Suppose we had

A_2 B_2,
A_2 B_1.

Because neither B_1 nor B_2 could be the parent species of the other, A_22 would have to arise below both of them (violating rule 3):

A_2 B_2

  A_22 B_21

A_2 B_1

  A_22 B_12

Pair 2 similarly requires that some species in the tree is A B_21, the arise point of B_21. Otherwise, B_21 would arise twice (violating rule 3). But the requirements from the two pairs are contradictory: one requires A B_21, the other requires A_22 B, and both should be ancestors of A_22 B_21. So either A B_21 should be derived from A_22 B or vice versa. But neither is possible by rule 1.

2. What would be a smallest possible set (there could be many) of species that cannot be derived from a tree where each species is described by two groups, two subscripts per group, and where each subscript can be either 1 or 2?

Here is a smallest possible set:

S1: A_11 B_11,

S2: A_11 B_21,

S3: A_22 B_21.

As in the solution to 1, the pair

S1: A_11 B_11,

S2: A_11 B_21

entails that A_11 B must be a species, and the pair

S1: A_11 B_21,

S3: A_22 B_21

entails that A B_21 must be a species. These cannot both be ancestors of A_11 B_21.

3. Find the nontree derivation of this set with the fewest interbreeding events:

S1: A_11 B_11,

S2: A_11 B_21,

S3: A_11 B_12,

S4: A_11 B_22,

S5: A_21 B_11,

S6: A_21 B_21,

S7: A_21 B_12,

S8: A_21 B_22,

S9: A_12 B_11,

S10: A_12 B_21,

S11: A_12 B_12,

S12: A_12 B_22,

S13: A_22 B_11,

S14: A_22 B_21,

S15: A_22 B_12,

S16: A_22 B_22.

This is every possible combination of groups. Initially we construct a tree:

```
A B
  A_1 B_1
    A_11 B_11
    A_12 B_12
  A_2 B_2
    A_21 B_21
    A_22 B_22
```

Every other set (all 12) can be derived from these giving 12 interbreeding events. I believe this is the minimum. (One could imagine interbreeding at the level above to produce A_1 B_2 and A_2 B_2, but this would violate rule 3 for their descendants. For example, A_11 would then arise twice.)

4. *The previous example had every possibility. Here is one with fewer possibilities. Are there different subsets of species which result in different trees?*

S1: A_11 B_11,
S2: A_11 B_21,
S3: A_11 B_12,
S4: A_11 B_22,
S5: A_21 B_11,
S6: A_21 B_21,
S7: A_21 B_12,
S8: A_21 B_22.

Yes. Here is one possibility:

```
A B
  A_1 B
    A_11 B
      A_11 B_1
         A_11 B_11
         A_11 B_12
      A_11 B_2
         A_11 B_21
         A_11 B_22
  A_2 B
    A_21 B
```

The other four can come from interbreeding.
Alternately, one can have:

```
A B
  A_1 B_1
    A_11 B_1
      A_11 B_11
      A_11 B_12
```

```
A_2 B_2
  A_21 B_2
    A_21 B_22
    A_21 B_21
```

This may be more pleasingly symmetric, but it still requires four interbreeding events. I don't know of any way to get fewer.

5. *Find a tree and two interbreeding events to derive:*

S1: A_11 B_11,

S2: A_11 B_21,

S3: A_11 B_12,

S4: A_11 B_22,

S5: A_21 B_11,

S6: A_21 B_12.

Starting with this tree,

```
A B
  A_1 B
    A_11 B
      A_11 B_1
        A_11 B_11
        A_11 B_12
      A_11 B_2
        A_11 B_21
        A_11 B_22
  A_2 B
    A_21 B
```

you need to add only two interbreeding events to produce A_21 B_11 and A_21 B_22. Not all trees are as good. For example,

```
A B
 A B_1
  A B_11
    A_1 B_11
    A_11 B_11
  A B_12
    A_2 B_12
    A_21 B_12
```

No others could be put in the tree. For example, A_11 B_21 (S2) can't be derived under a B_2, because then A_11 would arise twice. Similarly, A_21 B_11 (S5) can't be derived under A B_11 without violating rule 3.

6. *How many species could you describe with a tree having* k *groups, two subscripts, and two values per subscript?*

The answer is still just four. Here is why. For any pair of groups, the number is just four. Adding a group potentially adds constraints. It doesn't have to add constraints, however, because group C, for example, can have the same subscript as group A in every species.

S1: A_11 B_11 C_11,
S2: A_11 B_11 C_12,
S3: A_12 B_12 C_11,
S4: A_12 B_12 C_12.

7. *How about* k *groups,* m *subscripts, and* n *values per subscript?*

Just n^m. Here again the number of groups doesn't matter. All that matters is that each subscript can vary by n values.

A nontechnical introduction to graph-of-life style evolution in action can be found in the delightful book *The Beak of the Finch: A Story of Evolution in Our Time* by Jonathan Weiner (Vintage Books; Reprint edition, June 1995). It focuses on the Grant family's (the kids were involved, too) work on finches.

GRIDSPEED **SOLUTION**

1. How do you visit every intersection if you are forbidden from visiting any intersection twice?

If you can't visit any intersection twice, one good solution is shown in Figure S9. Start at (1,1). The points are: (1,1), (2,1), (2,2), (1,2), (1,3), (3,3), (3,1), (4,1), (4,4), (1,4), (1,5), (5,5), (5,1), (6,1), (6,6), and finally (1,6) at a cost of 12.1 hours.

2. Can you do better if you are allowed to visit intersections twice?

By allowing U-turns, it is possible to save 11 minutes more, bringing the time down to 11 hours, 55 minutes. The idea is to

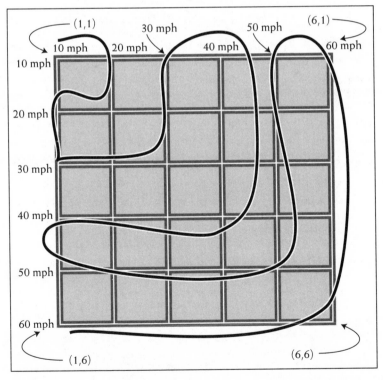

Figure S9 Fastest route to cover all grid points. This is a self-similar Z-like pattern.

retrace old paths in a clever way. Nico Ghilardi was the first to suggest this idea to me. His solution goes as follows. "The first 3 minutes can be gained on the way from (1,4) to (1,5) by following the edges (1,4), (2,4), (2,5), (1,5) instead of going directly. In a similar way it is faster to go around the block than directly from (5,1) to (6,1)—this saves 8 minutes."

Also, Richard Kandarian wrote a program to generalize this problem to larger grids. For example for a 10-by-10 grid, he finds that the fastest time without U-turns is nearly 26 hours whereas U-turns bring the time down to just under 24.5 hours. His website is at www.kandarian.com/page2.html.

3. *Is* (1,1) *the best place to start to visit every intersection in the shortest time possible with the no two-visit restriction?*

Jonathan "Air" Mason proposed the following proof of the optimality of starting at (1,1) when no U-turns are allowed. He observes that in any path, the vertices (the intersections) in the path will connect two edges (road segments) in the path unless they are one of the two end vertices. The end vertices connect to only one edge in the path. In the 12.1-hour solution, for each vertex, the two edges in the path that touch the vertex already use the fastest two choices for that vertex. Further, the first and last edges are the best choices for the first and last vertices. Therefore, the sum of the times required to travel across all of the intersections is the smallest possible. Also, notice that if we do not start (or end) at (1,1), then we will have to go through both edges incident with (1,1) at the expense of some other, faster edge, and making the whole path take at least 30 minutes longer. Thus, there are exactly four optimal solutions, all of which either start or end at (1,1). This solution can be generalized to square grids of arbitrary size.

VARIANTS.

H. V. Jagadish proposed the following variant of the puzzle: You are required to visit all intersections and return to the original starting point. He conjectures that on any square grid 3 by 3 or greater, you cannot do it without backtracking.

STACKING THE DECK **SOLUTION**

1. *Suppose that Bob arranges all cards with face value greater than four. Then Alice, without looking at what Bob has done, arranges the remaining cards and puts her cards after Bob's on the deck. Can she still force a win?*

If Bob can arrange the cards having values 5, 6, and 7, Alice can still win. Let Bob's cards be the x's. Every one of them will cause a round that will end with a card in the set that Alice has set up. All of those lead to victory for Alice: x, x, x, x, x, x, 4, 4, 3, 3, 2, 2, 1, 1.

2. *Suppose Bob takes seven cards and arranges them as follows: 1, 2, 2, 3, 3, 4, 4. Can Alice still win?*

Alice can still win by inserting cards as follows: 1, 2, 2, 3, 3, 4, [7, 6, 5,] 4, [7, 5, 6, 1].

3. *But can Alice win in general?*

By an elaborate case analysis that only a computer does willingly, one can prove that Alice will win even if Bob chooses seven cards, arranges them, and Alice preserves their order. The key insight is that Alice wants to arrange a consecutive string of six cards (or seven if starting at the very beginning) that will lead to a win for Alice. At the end of some round, one of those cards must be the last card dealt and so Alice will win. An elegant "closed form" proof of a few pages is still to be discovered.

ANALYTICAL LINEBACKING **SOLUTION**

1. *Can the runner or the tackles guarantee a win if the tackles can place themselves anywhere that is at least three spaces away from the runner and the runner starts on the top row?*

The tackles can force a win. The runner has the most freedom if he starts near the middle. The tackles respond by three next to one another three rows down:

```
. . R . . .
. . . . . .
. . . . . .
. T T T . .
. . . . . .
. . . . . .
```

The runner might move here:

```
. . . . . .
. . . . . .
. . . . R .
. T T T . .
. . . . . .
. . . . . .
```

The tackles could respond by retreating diagonally southward:

```
. . . . . .
. . . . . .
. . . . R .
. . . . . .
. . T T T .
. . . . . .
```

Even if the runner moves here

```
. . . . . .
. . . . . .
. . . . . .
. . . . . .
. . T T T R
. . . . . .
```

two tackles can block at the last row:

```
  .   .   .   .   .   .
  .   .   .   .   .   .
  .   .   .   .   .   .
  .   .   .   .   .   .
  .   .   T   .   .   R
  .   .   .   .   T   T
```

Any other second move by the runner can be blocked. In fact, the
tackles can always begin at that position.

With the same initial configuration

```
  .   .   R   .   .   .
  .   .   .   .   .   .
  .   .   .   .   .   .
  .   T   T   T   .   .
  .   .   .   .   .   .
  .   .   .   .   .   .
```

the runner could move here:

```
  .   .   .   .   .   .
  .   .   R   .   .   .
  .   .   .   .   .   .
  .   T   T   T   .   .
  .   .   .   .   .   .
  .   .   .   .   .   .
```

But then the tackles advance (and they would achieve this configuration for any other single move of the runner):

```
  .   .   .   .   .   .
  .   .   R   .   .   .
  .   T   T   T   .   .
  .   .   .   .   .   .
  .   .   .   .   .   .
  .   .   .   .   .   .
```

2. *If the tackles must start on the last row, does this change the outcome? Say how either way).*

The runner can force a win if the tackles start in the fifth or sixth row and the runner makes only one move in the first round. (This solution is primarily due to Gary Johnson and Mike Jarvis.) It goes like this:

```
. .  R . . .
. . . . . .
. . . . . .
. . . . . .
. . . . . .
. T T . . .
```

After the first move of both (if the tackles stay together, the runner can surely pass them):

```
. . . . . .
. .  R . . .
. . . . . .
. . . . . .
. T T . T .
. . . . . .
```

After the second move of the runner:

```
. . . . . .
. . . . . .
. . . . . .
. . . .  R .
. T T . T .
. . . . . .
```

The tackles cannot block this.

VENTURE BETS **SOLUTION**

1. Suppose that instead of asking for a 60 percent chance of attaining $60 million or more, you were willing to make that much with a likelihood of only about 34 percent. On the other hand, you want to reduce your chances of losing to under 5 percent. How would you do that?

You would spread your $12 million across all 11 companies. To make $60 million, you would need six winners. This happens only with probability 1/3—approximately. On the other hand, just two hot companies would be enough to ensure that you do better than break even. According to Luis Villalobos of Tech Coast Angels, every good venture capitalist would do this to reduce downside risk. In both the Villalobos solution and the one proposed in the warm-up, the expected value is the same. The warm-up solution I suggested achieves higher odds of a big win, but a bigger chance of losing too. The point is there is a choice.

2. Can you find a way to achieve your investors' goals (95 percent chance that their fund will grow from $17 million to $100 million), while keeping as much money as possible in reserve?

Spread your investments over 10 companies, giving each one $1.384 million, then the chances that at least 7 will win is over 0.95. If that happens the total amount including the reserve is a little over $100 million. $3.16 million is in reserve for future investments. Tim Ross contributed to the solution.

To figure out this probability, we use a small bit of theory. If the probability an individual company enjoys success is p, then the probability of exactly k successes out of n then is
$$(n \text{ choose } k)\, p^k\, (1 - p)^{(n - k)}$$
and the probability of at least k successes is the probability of k successes + probability of $k + 1$ successes + ... + probability of n successes. (n choose k is $n!/(n - k)!k!$ where $n!$ is $1 \times 2 \times 3 \times \ldots \times n$). The probability that at least seven will win is therefore 0.1298337 + 0.2758967 + 0.3474254 + 0.1968744, which is just over 0.95.

3. *What if you wanted to get a return of over $140 million but were willing to have a reserve of only $2 million. Then what could you do?*

If you spread your investments over six companies—giving each $2.5 million—then your chance of getting over $142.5 million is slightly over 95 percent. You'll have $2 million left in reserve. This solution is due to Narayan Ramanathan.

NOTES.

The reasoning underlying this problem applies to many different application domains. For example, a message network may try to increase the reliability of message delivery by sending the message through many routes. It is in that context that I first heard of the basic mathematics of the problem from Aris Tsirigos's master's thesis at Cornell.

BLUFFHEAD **SOLUTION**

1. *Caroline says, "I don't know." David says, "I don't win." Then Jordan says, "I win."*

As in the warm-up, Caroline must not see an ace. David sees no ace either so he knows that king is high. David must therefore see a king on Jordan or Caroline's foreheads. Jordan knows this but sees no king on Caroline's or David's foreheads so declares himself the winner. Conclusion: Jordan has a king. Both Caroline and David have less.

2. *Caroline says, "I don't know." David says, "I don't win." Then Jordan says, "I tie."*

As in question 1, the king is high. David must see a king somewhere. Jordan sees that Caroline has no king but David does, so Jordan declares that he is tied for a winner. Note that if Caroline had had a king, then Jordan would have said, "I don't win."

3. *Caroline says, "I don't know." David says, "I don't win." Then Jordan says, "I don't win."*

As in questions 1 and 2, Caroline sees no ace. David must see a king somewhere so he says he doesn't win. Jordan sees that Caroline has a king, so Jordan knows he doesn't win. That is why he says he doesn't win. (If Caroline sees no king, then she will claim victory in the next round.)

4. *Caroline says, "I don't know." David says, "I don't know." Then Jordan says, "I don't know." Then (having heard this) Caroline says, "I lose." What do David and Jordan then say and what do you know about the cards?*

Caroline cannot see an ace or she would know that she could not win. David cannot see an ace or king or he would know that he could not win. Jordan cannot see an ace, king, or queen or he would know that he could not win. Caroline must see a queen. It must be on Jordan otherwise Jordan would have seen it and said he doesn't win. David knows this, so he will say "I lose" and Jordan will say "I win."

5. *Caroline, David, and Jordan say, "I don't know" in order, then say, "I don't know" in a second round. In the third round, Caroline and David say, "I don't know," but Jordan says he wins. What does Jordan have and what might he see?*

Caroline must have a five and Jordan a six. David may have a four or a five. Here is why. Eight "don't knows" in a row require some amount of systematic approach to our analysis. To eliminate the top set is straightforward. Caroline's saying "don't know" the first time means neither David nor Jordan has an ace. David's saying "don't know" next means neither Caroline nor Jordan has a king or higher. Jordan saying "don't know" next means neither Caroline nor David has a queen or higher. This continues, until after the eighth "don't know," we know that Caroline and Jordan both have a six or lower, and David has a seven or lower.

The lower cards are slightly more complicated. Caroline's saying "don't know" first means that *,2,2 is eliminated. David's saying "don't know" means that 2,*,2 and 3,*,2 are eliminated. 3,*,2 is eliminated because otherwise David would know he at least tied. Jordan's saying "don't know" eliminates all cases where Caroline and David both have three or less. That's the first round.

In the second round, Caroline's saying "don't know" again eliminates all the cases where David has a three or less and Jordan has a four or less, as well as the cases where David has a four and Jordan has a two. David's saying "don't know" then eliminates all the cases where both Caroline and Jordan have a four or less *and* the cases where Jordan has a two and Caroline has a five or less. Jordan's saying "don't know" then eliminates the cases where both Caroline and David have a four or less, and also the cases where Caroline has a five and David has a three or less.

In the third round, the analysis becomes more complicated, but after the next two "don't knows," the only remaining cases imply that Caroline holds a five or a six, David holds a four, five, six, or seven, and Jordan holds a six; all eight cases are possible at this point. Jordan's saying "win" indicates that the other two must hold cards less than a six, so Caroline must hold a five, and David might hold a four or a five.

NOTES.

Raymond Smullyan wrote a two-player version with some similar reasoning called smullyan/integers at http://einstein.et.tudelft.nl/~arlet/puzzles/logic.html.

An important source of inspiration, however, comes from the work on the logic of knowledge, where one reasons not only from what one knows but also from what others know or don't know. Joe Halpern, currently at Cornell, has served as the spearhead (or lightning rod) for recent work in this area with many papers and two very accessible books that he has co-authored: *Reasoning about Knowledge* and *Reasoning about Uncertainty*.

My thanks to Tom Rokicki for correcting an earlier solution to question 5 and verifying the other answers. He has constructed a nice website with his further detailed analysis: http://tomas.rokicki.com/bluffhead.

ADVERSARIAL BIFURCATIONS
SOLUTION

Can you do as well or better?

Here is the solution for Baskerhound's tree:

A sc0:5219 sc1:2055 sc2:396
 B sc0:4894 sc1:1776 sc2:-2408
  D sc0:4894 sc1:1055 sc2:-3092
   H sc0:3791 sc1:1055 sc2:775
    P sc0:1055 sc1:775 sc2:775
    Q sc0:3791 sc1:3011 sc2:3011
   I sc0:4894 sc1:-3092 sc2:-4077
    R sc0:-3092 sc1:-4077 sc2:-4077
    S sc0:4894 sc1:-465 sc2:-465
  E sc0:3046 sc1:1776 sc2:-2408
   J sc0:3046 sc1:1776 sc2:-2408
    T sc0:2149 sc1:1776 sc2:1776
    U sc0:3046 sc1:-2408 sc2:-2408
   K sc0:2523 sc1:-1043 sc2:-3266
    V sc0:18 sc1:-1043 sc2:-1043
    W sc0:2523 sc1:-3266 sc2:-3266

C sc0:5219 sc1:2055 sc2:396
　F sc0:5219 sc1:2055 sc2:396
　　L sc0:2189 sc1:452 sc2:396
　　　X sc0:452 sc1:396 sc2:396
　　　Y sc0:2189 sc1:483 sc2:483
　　M sc0:5219 sc1:2055 sc2:-627
　　　Z sc0:5219 sc1:-627 sc2:-627
　　　a sc0:3247 sc1:2055 sc2:2055
G sc0:4125 sc1:509 sc2:-6
　N sc0:509 sc1:115 sc2:-3841
　　b sc0:115 sc1:-3841 sc2:-3841
　　c sc0:509 sc1:187 sc2:187
　O sc0:4125 sc1:667 sc2:-6
　　d sc0:4125 sc1:667 sc2:667
　　e sc0:2338 sc1:-6 sc2:-6

The notation means the following. sc0 is the score you can get from the current node if your adversary has 0 more moves to make, sc1 if your adversary has 1 more move to make, and sc2 if your adversary has 2 more moves to make. These numbers are derived in a bottom-up fashion. The basic method is this. The 0 score for a node is the maximum of the 0 scores for its children. The 1 score for a node is the minimum of the 0 scores for its children (if the adversary chooses now) and the maximum 1 score for its children (if the adversary refrains from choosing). Similarly, the 2 score for a node is the minimum of the 1 scores for its children (if the adversary chooses now) and the maximum of the 2 scores for its children (if the adversary refrains from choosing).

NOTE.

If you found the Delphi Flip puzzle difficult, try it again now. In fact you could try situations where the oracle could lie several times and yet you would still win provided the oracle told you the truth sufficiently more often.

ULTIMATE TIC-TAC-TOE **SOLUTION**

Is there a winning strategy for X from the shown position?

Define a corner to be the three perimeter cells nearest a corner. X must have two out of three cells in at least one corner, since there are only six O's in the entire perimeter (see Figure S10). X can force a win by choosing a corner in which X has a majority and placing an X just inside that corner. This creates two possibilities for X to win a threesome in the final move.

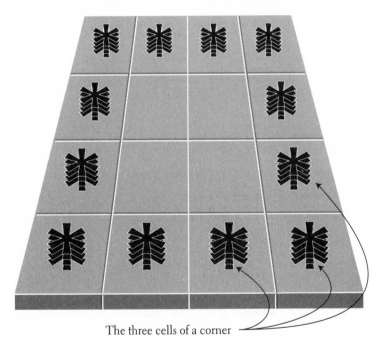

The three cells of a corner

Figure S10 Solution: X must be able to force a win because there must be an X in two out of three of the cells in some corner (otherwise O would have moved eight times just in the perimeter, where O has only six moves.)

The winning move consists of placing an X nearest some corner where there are two X's. This ensures that X cannot be blocked from a threesome in his final move.

JUMP-SNATCH **SOLUTION**

1. How many spaces must you leave empty and where should they be in order to have only one piece remaining at the end?

For the 3-by-3 square, leaving three spaces empty works:

X	X	.
X	X	X
.	X	.

Take the piece in the upper-left-hand corner and go all around and then through the diagonal.

2. Can you prove that this is the minimum?

Three empty spaces are the minimum, because no piece on the corner can get jumped, so at least three corners must be empty.

3. If the tic-tac-toe board were 4 by 4, how many spaces must you leave empty and which one(s) to have only one remaining piece at the end?

For the 4-by-4 square, you can start with one empty space and guarantee to have only one piece standing at the end.

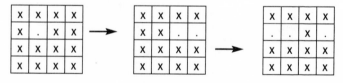

Now do the first, second, and fourth *columns* just as the row above was done.

.	.	X	.
.	.	X	.
X	X	X	X
.	.	X	.

Now jump up and left diagonally.

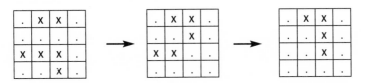

Now, go back and forth with the upper-left piece and end as follows.

.	.	.	.
.	.	.	.
.	.	.	X
.	.	.	.

4. *Who wins the displayed game?*

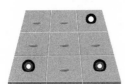

Jump-Snatch V, VI, VII: Jumper jumps up on right side. Then Snatcher jumps right on bottom. Then Jumper jumps down on left.

Jump-Snatch VIII: Snatcher can't jump so has three pieces to move to the center. Snatcher wisely chooses to create a diagonal.

Jump-Snatch IX: Jumper must move a noncenter piece.

The Snatcher wins. The next three moves are forced. Then the Snatcher moves to the center and Jumper is lost.

SHORT TAPS **SOLUTION**

1. Can you show how to do it in 15 minutes or less, again assuming that you start messages on the minute?

Here is a 14-minute solution for the seven-message case:

A and B and F start at 0;

G starts at 3;

D starts at 6;

C starts at 7;

E starts at 8.

The entire transmission is over at time 14.

2. If in addition to these seven messages you had to send three four-minute messages, then can you do it in 20 minutes or less without allowing more than three messages to be tapped in their entirety and under the on-the-minute assumption?

Let the additional three four-minute messages be labeled H1, H2, and H3, then we can finish in 18 minutes as follows:

A and B and H1 start at 0;

F starts at 4;

D starts at 6;

G and H2 start at 7;

E starts at 11;

F and H3 start at 14.

3. To what extent can you improve your timing if you are willing to start sending messages without insisting that they be on the minute?

We can reduce the total transmission time by nearly 2 minutes in both cases. For the first version of the problem:

D and E and F start at 0 minutes, 0 seconds;

G starts at 2 minutes, 1 second;

A starts at 8 minutes, 1 second;

C starts at 8 minutes, 2 seconds;

B starts at 9 minutes, 2 seconds.

Note that G, for example, must start at slightly more than 2 minutes, a nanosecond more would be enough, to avoid finishing within the first 10 minutes. We use the term *second* only for concreteness.

For a 15+ minutes solution to the case where there were three extra four-minute messages (H1, H2, H3):

A and C and H1 start at time 0;

F starts at time 3 minutes, 1 second;

E starts at time 4 minutes, 1 second;

D starts at time 5 minutes, 1 second;

G starts at time 5 minutes, 1.5 seconds;

H2 and H3 start at 11 minutes, 2 seconds;

B starts at 12 minutes, 2 seconds.

This solution is due to Walt Stadlin.

GRAB IT IF ƳOU CAN **SOLUTION**

1. What should the plaintiff ask for to maximize his expected receipts in the case that he suspects the defendant can read what is in the plaintiff's sealed envelope?

This one requires either hacking or a bit of calculus.

Let $3 million be L (standing for low) and $10 million be H (high). Suppose the plaintiff asks for P and the defendant offers D. It is reasonable to assume that D will be lower than P, because the defendant has no rational reason to give a gift to the plaintiff. Also, the defendant loses out if he falls far below L (that is, below $L - (H - L)$) because then the judge chooses between L and H and the plaintiff can ask for and receive H.

The midpoint between P and D is $M = (P + D)/2$. By the above considerations, we suppose $M \geq L$. Because the judge decides on a dollar amount between L and H with uniform probability, the plaintiff's probability of receiving P is $(H - M)/(H - L)$. Otherwise he receives D. So, the plaintiff's expected gain in this case is the weighted average of receiving P and receiving L:

Payment = $(P \times (H - M)/(H - L)) + (D \times (M - L)/(H - L))$

A constructive method to find the best solution is to choose each plaintiff value and assume that the defendant does the worst thing possible (from the plaintiff's point of view). For example, the worst that the defendant can do if the plaintiff chooses $3 million is to choose three (or less). So, in every case, we assume the defendant is as clever as possible. The maximum of these worst-case situations occurs when the plaintiff asks for 10 (the high value). In that case, the defendant should choose three. The expected gain is $6.5 million.

We can also do this by using calculus:

Payment
$= (P \times (H - P/2 - D/2))/(H - L) + (D \times (P/2 + D/2 - L))/(H - L)$
$= (PH - PP/2 - PD/2)/(H - L) + (DP/2 + DD/2 - LD)/(H - L)$
$= (PH - PP/2 + DD/2 - LD)/(H - L)$

Differentiating this with respect to P (D is fixed but unknown) and setting it equal to 0, we get $0 = H - P$. So $P = H$.

So, the plaintiff should always ask for the highest amount in the range regardless of whether the defendant knows what the plaintiff will ask. Carl Bosley suggested this relatively simple calculus approach.

2. *How will the plaintiff's request change if he knows that his sealed envelope is secure?*

It doesn't change at all.

FURTHER READING.

Sze Shing Wong and Paul Fage observed that this problem can be viewed as an application of Nash equilibrium. In that formulation, the plaintiff tries to maximize his/her payoff given the distribution of the judge's decision possibilities and the range of values that D can take. The defendant tries to minimize his/her payout. The Nash equilibrium is the optimal strategy pair for both parties. Because the distributions may not be uniform, the optimal choices for the defendant and the plaintiff might then be different. As we saw in the warm-up, if the judge were extremely likely to give a number N, then the optimal defendant and plaintiff choices would cluster closely around N.

James Ryan pointed out that a similar scheme is used to arbitrate salaries in professional baseball. Normally, it brings the negotiating sides closer together, however. The reason for the difference could be that the arbitrator gives stronger valuation signals (that is, farther from uniform) in that case. Another consideration is that each side wants to do well even in its worst case rather than maximize its expected value.

SKYCHASERS **SOLUTION**

1. Can you determine how many containers flow from each source to each destination?

B to A: 1
A to B: 1

These two must hold because nothing is coming from the right side. Therefore three are going from A to right and one from B. None of those four can go to C because then the two that come up from the bottom would have nowhere to go. That's also why D to E and E to D are both impossible.

E to C: 1
D to C: 1
C to E: 2

The six going to the lower right consist of three from A, one from B, and two from C. Therefore, D gets at least one from A and possibly three. E gets at most two from A but possibly zero. The total between the two is three.

2. *Can you say how?*

Three or fewer can go from A to D and two or fewer from A to E. This form of traffic analysis is very imprecise. Louis Tagliaferro contributed to the solutions to this puzzle.

NANOMUNCHERS **SOLUTION**

1. Are there any shapes having the property that even an arbitrarily large graph of that shape has a solution regardless of the start node and regardless of the order of the loop?

An enormous single square having any number of grid cells is munchable regardless of the loop order and of the starting point.

2. Are there graphs that are munchable with two nanomunchers (with different starting points and possibly different loop orderings) but not with one?

The T shape that Liane identified is munchable with two nanomunchers but not with one.

3. Can you munch the graph in the figure?

Yes, start at the node labeled 1 and loop right, left, down, up:

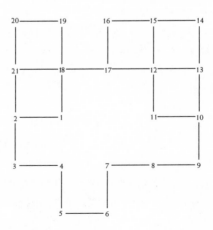

4. Can you munch a 4-by-4 grid with one nanomuncher?

A 4-by-4 grid can be handled by starting at position $(1,1)$ where the origin point is $(0,0)$. That gives a graph like this:

```
.  .  .  .

.  .  .  .

.  1  .  .

.  .  .  .
```

The first four moves assuming an up-right-left-down pattern color the entire center, leaving a closed ring that is surely munchable.

```
.  .  .  .

.  2  3  .

.  1  4  .

.  .  .  .
```

David Bunde and Rodney Meyer showed how to munch this grid and a few others.

5. Which n-by-m grids are munchable by a single nanomuncher?

Here are some munchable grids. Consider a grid of the form *n* by 1, *n* by 2, or *n* by 3 for any *n* that is mostly horizontal. That is, *n* by 2 or 3. Then one way to munch the grid is to start at the lower left and loop as follows: if *n* is even, then left, up, right, down; if *n* is odd, then up, right, down, left.

If the grid is *n* by *m* and both *n* and *m* are greater than three and odd, then start at the bottom left corner and use the sequence: up, right, down, left. You can make it all the way around. You will handle the bottom two rows, then the right side two, then the top two, then the left two and will start on the third from bottom row moving right.

STRATEGIC BULLYING **SOLUTION**

1. What is the largest set of distinct positive whole number values that is stable and whose greatest power is 21?

If the highest power is 21, then one large set of powers is 1, 2, 3, 4, 5, 6, 7, 8, 9, 15, 19, 20, 21 as pointed out by Travis Fisher of Penn State University. I know of no larger solution.

2. What if the positive whole number values need not be different?

If the whole number values need not be different, then there could certainly be twenty-one 1's and one 21.

Philip Gladstone of Framingham, Massachusetts, was the first to point out that, under certain assumptions, one can construct an arbitrarily large stable set if duplicates are allowed: all individuals have the same power and the number of entities is even. This is stable if everyone is risk-averse. To see why this is important, consider a case where there are four people with the same value V. If three gang up against the fourth, then one of those three will also certainly be destroyed because three people with value V are certainly unstable. So, if everyone is risk-averse, this will be stable.

3. Can you show a configuration that is stable when all entities are risk-averse but not when they are all risk-ready?

Consider, for example, X, X, X, X where X represents a power value. Three of the Xs could gang up on the fourth, leaving X, X, X. But then any two in coalition would defeat the other. So, each entity runs a risk in allowing the first attack to happen. In the risk-averse model, therefore, X, X, X, X is stable. In the risk-ready model, it isn't because each X believes that it might survive.

4. Can you prove that a stable risk-ready configuration will always lead to a stable risk-averse configuration? Or can you show this is false by giving a counterexample?

It's not true. A counterexample arises because other entities may understand these incentives and play to them. Consider 5, 5, 5, 14, 1. In a risk-averse environment, the 5's never want to be left alone or conquer the 14, because then one 5 will be destroyed. However, risk-averse 5's are willing to destroy the 1, because then the remaining configuration is stable. They won't destroy the 14 as argued above and they won't allow any 5 to be destroyed (because then the 14 could destroy everyone else). Risk-averse 14 can allow 1 to be destroyed. On the other hand, if everyone is risk-ready, then the three 5's are willing to be left alone. 14 and 1 both know this, so they will defend one another. The 5's will also defend one another because if any of them is destroyed, then 14 will conquer all of them.

5. *What about if some entities are risk-averse and some are risk-ready? Does either have a consistent advantage in terms of survival?*

Risk-ready would be better in the case X, X, X, X. If three X's are risk-ready, then they will conquer the remaining X and then two out of three of them will survive. On the other hand, if two are risk-ready and two are risk-averse, then all four will survive.

6. *Can there be several stable subsets of an unstable configuration?*

Yes. Consider 5, 6, 7, 8. Both 5, 6, 7 and 6, 7, 8 and others are stable.

COMMENTARY.

First a comment about the open problem. Stability means that every conflict will be stopped by a coalition interested in survival. This is certainly true in the base case of invincible configurations. Suppose our three rules of stability and instability hold up to size N. Consider configurations of size $N + 1$. Suppose that a stable configuration C has a battle that someone wants to fight that will eliminate some X. This implies that if there is a stable coalition C' that is a subset of C, then C' does not dominate C. Therefore if $C - X$ is more powerful than X, then $C - X$ must be unstable, hence some members of $C - X$

would then also be destroyed. These would all ally with X. Does this argument hold water?

The theory as it stands does not deal with the fact that a war costs a lot to the victors as well as the losers. But whether this makes them more or less powerful is not certain. Britain and France both became more powerful as a result of the wars they fought against one another until the early nineteenth century. When the police allow one street gang to dominate all others, the gang left is truly formidible. So, war can be destructive to the victors but may give them good practice for the next war.

A second open problem has to do with the uncertainty of power relationships. Sometimes the apparently weaker opponent wins, often because the stronger one lacks commitment. North Vietnam was weaker than the United States in the 1960s, Eastern Europe was weaker than the Soviet Union in the late 1980s, yet they both prevailed.

Uncertainty often leads to over-optimism. If both sides think they will win easily, they may go to war. World War I may be such an example. Highly risk-averse powers may therefore wait until they have twice the power of their opponents, a far more certain situation.

After seeing the deadly effects of the atomic bomb, the French developed a strategic doctrine they called "la force de frappe." Their metaphor was the bee sting. A bee protects itself not by overpowering its larger attackers, but because it can sting them. The sting will cost the bee its life, but is painful to the would-be aggressor. So the aggressor will not attack. France developed enough nuclear bombs not to destroy its enemy but enough to inflict unacceptable pain.

In our model, even weak entities can acquire la force de frappe: defensive force sufficient to destroy or inflict unacceptable damage to a superior attacking power though they will be destroyed themselves. For example, if a 5 attacks a 4 having a bee sting, then both the 5 and the 4 will be destroyed. So, 5 won't attack. The 4 with bee sting cannot, however, attack a 5 and prevail.

La force de frappe may seem inherently stabilizing, but is this always so? That is, might the acquisition of bee-sting capability render a stable setting unstable? Think before you read the last paragraph.

La force de frappe is not necessarily stabilizing. 5, 4, 3 is stable. 5, 4*, 3* is unstable (where an asterisk indicates a power having la force de frappe), because 3 and 4 can attack 5 with impunity.

THE STONE TOMB OF ZIMBABWE SOLUTION

1. How many rays do you need if you must set up all the rays before turning on any laser and you must be able to identify the locations of the corridors for sure?

Here is a solution with six rays:

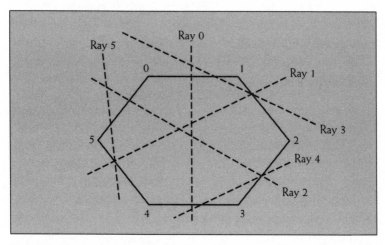

Figure S11 These six rays are enough (though not any six rays are enough).

2. Can you generalize this solution to higher n-gons (n > 6)? [hard]

Tom Rokicki, Michael Birken, Andrew Palfreyman, and Bruce Moskowitz sent in clever solutions to the puzzle. Under the assumption that all beams have to be placed before any tests are run, Rokicki was able to show that for any *n*-sided polygon where $n \geq 6$, there must be ceil($n \times (n - 4)/2$) lasers (where ceil(x) is the ceiling function that returns x if x is an integer and the next integer greater than x otherwise). Note that this number is close to the requirement that there must be a laser between every wall and every other wall.

Rokicki argues as follows for the case when n is large: "Consider two corners on the polygon that are not near each other (farther than two units away). Call their locations (a) and (b) where we number the corners sequentially around the polygon. Whatever laser combinations we set up must be able to distinguish the following two cases:

(a $\longrightarrow$ b + 1), (a + 1 $\longrightarrow$ b) (corridors are parallel)

vs

(a $\longrightarrow$ b), (a + 1 $\longrightarrow$ b + 1) (corridors cross)

The only laser firing that can make this distinction is one that fires from wall (a — a+1) to wall (b — b+1); all other lasers will always cross the same number of corridors."

So, the number of laser beams needed becomes proportional to the square of the number of segments.

3. If you could get feedback from previous tests before posing new ones, how many setups would you need?

Five setups are enough in this dynamic case. Start with (0 — 1, 1 — 2), (0 — 1, 2 — 3), (0 — 1, 3 — 4). (Here is the notation: the first setup fires a beam from between 0 and 1 to between 1 and 2.) After this, use case analysis to see that you can always find the unique answer after firing five beams. This approach was suggested by Bruce Moskowitz.

A SMALL HISTORICAL NOTE.

Centuries before Europeans arrived in Zimbabwe, the people there built cities out of stone. Tombs like the ones described in this puzzle may well exist.

PROTEIN CHIME **SOLUTION**

Can you create a "protein chime" that will ring every 70 seconds?

Each circuit will cycle periodically between a repressing mode and a nonrepressing mode. You have to arrange for the nonrepressing modes to coincide only once every 70 seconds.

A nice solution (suggested by Juan Carlos Campos) is to force C1 up for five seconds; force D2 up for five seconds; force E1 up for seven seconds; and force F2 up for seven seconds. This will have the result: C1 up from seconds 1–5; down from seconds 6–10; etc. D1 up from seconds 5–9; down from seconds 10–15; etc. E1 up from seconds 1–7; down from seconds 8–14; etc. F1 up from seconds 7–13; down from seconds 14–20; etc. T will chime at 71, 141, 211, 281, 351, 421, 491, 561, etc.

COMMENTARY.

These circuits each have an odd number of proteins. In this way, they will go up and down. One can also create a circuit that is stably up or stably down by forming a cycle having an even number of circuit elements. Timothy Gardner and James Collins of Boston University call this a "toggle switch."

LOST HIKER **SOLUTION**

1. Can you find the best planar shape and the best route on that shape?

As you can see from Figure S12 it is possible to construct a 100-square-mile area by attaching a semicircle to a rectangle as follows: make the rectangle 4 by $25 - \pi/2$ giving an area of $100 - 2\pi$. The semicircle of radius 2 has an area of 2π. Now the jeep needs to travel only $25 - \pi/2$ miles, which is approximately 23.43 miles. So the total time is about 234.3 minutes. This construction was suggested by Andrei Broder.

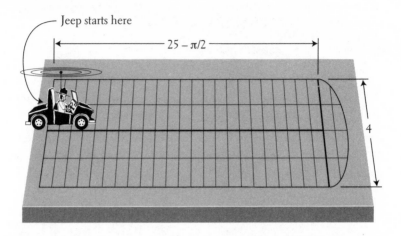

Figure S12 A sub-240 minute solution to the missing hiker problem given the best possible area to search. That area is a rectangle 4 by $25 - \pi/2$ miles in dimension and a semicircle with a radius 2. The jeep needs only to travel the length of the $25 - \pi/2$ rectangle.

2. *What is the best route you can find through the 10-by-10 square?*

The best known route on the 10-by-10 square yields an overall length of 29.085 miles. One can use the solver facility in Excel to obtain this (but I admit the math is not elementary).

The route begins on the left side:

1	0	2.659988899
2	0.765510457	1.847698985
3	5.906915717	2
4	8.584479676	1.412905758
5	8	4.065277393
6	8	6.104103112
7	8.540270859	8.632817808
8	5.932548019	8.251930006
9	1.404482698	8.57612281
10	1.558862621	5.912964019
11	4.351633644	5.128765987

This has a total length of 29.08507576 miles. The route is shown in Figure S13. This solution is due to Matthew Self and Ray Tomlinson building on work by Robert Dawson, and Martin and Erik Demaine.

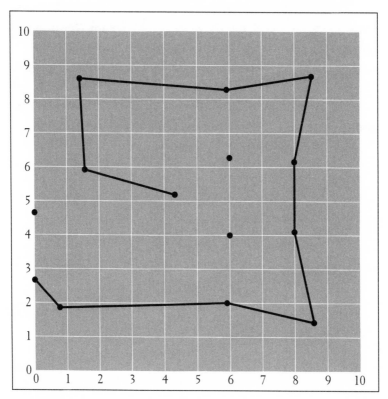

Figure S13

3. *What if, to conserve energy, the distress signal goes on for one second every 10 minutes? What would be the best route then?*

When the distress signal is off most of the time, the searchers should be within radio range of each point for as long as the off portion of the signal (the length of the on period doesn't matter). For example, if the distress signal can be off for up to 10 minutes, then the Jeep can travel a mile while the distress signal is off. To handle this, the jeep sweeps out effective paths that are $\sqrt{2^2 - (1/2)^2} = 1.94$ miles wide. This is the height of an isosceles triangle with base of one mile and sides of two miles.

Extra care is also needed in the corners. Since the searchers don't know when the signal will be on or off, they must remain within range of the corners for 10 minutes to be sure the signal will be heard. During the 10-minute wait, the Jeep can continue to move as long as it remains within two miles of the corner.

The endpoint of the path is exactly two miles from the furthest points that were not reached in the first pass (A and B), but, again, the Jeep must wait there 10 minutes to be sure the signal is heard.

Similarly, at the start of the path it takes a full 10 minutes to establish point C (the point farthest up the west edge that is known to have been checked). Even though the distance from the edge to the first turn is only 0.53 miles, the Jeep must wait there until 10 minutes have passed before proceeding southward.

The total distance of the path is 31.894 miles and the total time is 333.65 minutes.

Here are the detailed data for the 10-minute case:

X	Y	Distance	Time
0.000	2.334	0.000	0.000
0.529	2.329	0.529	10.000
0.645	1.893	0.451	4.513
1.481	1.344	1.000	10.000
5.428	1.936	3.992	39.915
6.419	1.798	1.000	10.000
8.266	0.996	2.013	20.134
8.965	1.711	1.000	10.000
8.171	3.495	1.953	19.527
8.064	4.489	1.000	10.000
8.064	5.624	1.135	11.348
8.219	6.612	1.000	10.000

8.944	8.301	1.838	18.384
8.254	9.025	1.000	10.000
6.502	8.400	1.860	18.599
5.509	8.285	1.000	10.000
1.662	8.888	3.893	38.932
0.916	8.222	1.000	10.000
1.013	5.982	2.242	22.418
1.721	5.276	1.000	10.000
4.705	5.123	2.988	29.880
4.705	5.123	0.000	10.000
Total		31.894	333.651

Auxiliary points:

A	6.192	3.785
B	6.228	6.419
C	0.000	4.258

This solution is due to Matthew Self.

NOAH'S ARC **SOLUTION**

1. What is the shortest sequence S you can imagine such that even if strands can be six characters long, you won't be able to verify S?

Suppose S were AAAAAATAAAAAA. With strands of length 6, there would be no way to distinguish between S and S' = AAAAAAATAAAAA. Both S and S' have the same number of A's, the same number of T's and no combination of A's and T's of length 6 can distinguish one from the other.

2. If S = ACGAC, then two strands of length 2 are enough. Can you see how?

If S = ACGAC, then test with AC and GA. The AC will display in two colors. The GA will display in one. You already know how many of each kind of nucleotide there are. So this distinguishes ACACG from ACGAC. This was suggested by Andrew Palfreyman.

3. *What if we have AC repeated N times?*

AC repeated N times can be matched by AC, AA, CC. AC matches both the first AC and the interior ones. AA matches nowhere, nor does CC. Therefore, A and C must alternate.

4. *Here is the sequence we want to verify:*

TCACTCGGCTCTCGCACACGGAGATAGCTC.
What is the smallest size and the smallest number of strands of that size that would verify this sequence?

Liane was able to use 10 strands of length 6 each:

TCACTC ACTCGG   TCGGCT GGCTCT TCTCGC
CGCACA CACGGA   GGAGAT ATAGCT TAGCTC

LIQUID SWITCHBOARD **SOLUTION**

Is it possible to get any permutation of the five colors using only three levels of switches?

The following permutations are not possible with three levels of switches though they are for four levels:

B, C, D, E, A;
E, A, B, C, D.

For example, to get the first one:

1. Swap A with B in the first level to get B, A, C, D, E.
2. Swap A with C in the second level to get B, C, A, D, E.
3. Swap B with C and A with D in the third level to get C, B, D, A, E.
4. Swap C with B and A with E in the fourth level to get B, C, D, E, A, as desired.

PRIME SQUARES **SOLUTION**

1. Can you find an omnidextrous prime 3-square that uses as few distinct digits as possible?

Here is an omnidextrous prime 3-square that uses only three different digits:

3	1	1
1	8	1
1	1	3

How might you have figured out that three digits are necessary for the omnidextrous prime 3-square? A two-digit solution isn't possible because any such combination cannot involve an even digit, 5, or 0. (Such a number could only be placed in the center; if it were placed on the side, the three digit number ending with that digit would not be a prime. But if the even number were placed in the center the numbers on the top and bottom would be the same.) Each remaining combination of odd numbers has at least one combination that is not prime, for example, 133, 119, 717, 737, 933, and 779. For this reason, it is not possible to make an omnidextrous prime 3-square out of two digits.

2. How about a prime 5-square that uses all 10 digits?

Here it is:

1	6	4	5	1
4	5	3	8	9
9	2	8	9	3
2	8	0	6	9
9	7	1	7	1

REPELLANOIDS **SOLUTION**

1. What is the smallest circumference of the cylinder in the full problem?

green, crimson = 8, 6;

aqua, blue, blue = 4, 5, 5;

daisy, daisy = 7, 7;

crimson, aqua, aqua = 6, 4, 4.

2. A second research group has discovered a second virus called a mini-repellanoid. Here is what they know: every strand has length 2 or greater, there are five differ-ent strand sizes, the circumference is 7, and like repels like at a distance of 3. How could this be?

For mini-repellanoids: let the strand lengths be $A = 2$, $B = 3$, $C = 4$, $D = 5$, $E = 7$. The circumference can then be 7. Here is how: E, DA, AAB, BC.

SAFECRACKING **SOLUTION**

How can you find the combination of a safe having 10 switches, each of which has three values?

Just 15 settings are sufficient. The X represents "don't care." So, you can leave the switch in the low position if an X is written. A represents low, B middle, and C high.

Number	S1	S2	S3	S4	S5	S6	S7	S8	S9	S10
1	A	A	A	A	A	A	A	A	A	A
2	A	B	B	B	B	B	B	B	B	B
3	A	C	C	C	C	C	C	C	C	C
4	B	A	B	C	A	B	C	A	B	C
5	B	B	C	A	B	C	A	B	C	A
6	B	C	A	B	C	A	B	C	A	B
7	C	A	C	B	A	C	B	A	C	B
8	C	B	A	C	B	A	C	B	A	C

9	C	C	B	A	C	B	A	C	B	A
10	X	A	A	A	B	B	B	C	C	C
11	X	A	A	A	C	C	C	B	B	B
12	X	B	B	B	A	A	A	C	C	C
13	X	B	B	B	C	C	C	A	A	A
14	X	C	C	C	A	A	A	B	B	B
15	X	C	C	C	B	B	B	A	A	A

Yuichi Tanaka was the first to point this out.

MATHEMATICAL NOTES.

Wim Caspers, a theoretical physicist, observes that this class of problem can sometimes be solved nicely by the use of subgroup generators. For example, if we equate A with 0, B with 1, and C with 2, then we can start off the solution to the warm-up question with all 0's: 0, 0, 0, 0 and apply a pair of transformations. One of the pairs he suggested was 1, 2, 2, 1 and 0, 1, 2, 2.

By applying a pair of transformations, we mean the following. We apply the first transformation until we are about to repeat ourselves, then we apply the second transformation to the state that was to be repeated until we cannot do anything else without repeating a previous configuration. Applying means adding with a modulus.

Starting with 0, 0, 0, 0 we apply 1, 2, 2, 1 which means adding in those values at their corresponding positions, so the next setting is 1, 2, 2, 1. We do this again and get 1, 2, 2, 1 + 1, 2, 2, 1 = 2, 1, 1, 2. Note that the second position yields 1 because 2 + 2 = 1 in this case and addition here is modulo 3 (the only possible values are 0, 1, and 2). Continuing, if we do 2, 1, 1, 2 + 1, 2, 2, 1, we get 0, 0, 0, 0. This is a repeat, so it is time to apply the second transformation on the repeated value of 0, 0, 0, 0. So, the next entry is 0, 0, 0, 0 + 0, 1, 2, 2 = 0, 1, 2, 2.

Now we go back to using the first transformation, giving us 0, 1, 2, 2 + 1, 2, 2, 1 = 1, 0, 1, 0. Repeating this addition we get 1, 0, 1, 0 + 1, 2, 2, 1 = 2, 2, 0, 1. Then 2, 2, 0, 1 + 1, 2, 2, 1 = 0, 1, 2, 2 but that is a repeat. So, we take the repeated value 0, 1, 2, 2 and add in the second transformation, yielding 0, 1, 2, 2 + 0, 1, 2, 2 = 0, 2, 1, 1.

Now we go back to using the first transformation, giving us 0, 2, 1, 1 + 1, 2, 2, 1 = 1, 1, 0, 2 and then finally 2, 0, 2, 0.

The final result is one solution for the warm-up (but using numbers instead of letters):

0, 0, 0, 0

1, 2, 2, 1

2, 1, 1, 2

0, 1, 2, 2

1, 0, 1, 0

2, 2, 0, 1

0, 2, 1, 1

1, 1, 0, 2

2, 0, 2, 0

The motivation for choosing such a pair of transformations is the following: for each pair of input positions we want the corresponding pair of transformation positions to generate all possible combinations. For the transformation we've used (1, 2, 2, 1; 0, 1, 2, 2), consider the first and second positions of each transformation: (1,2) and (0,1). Starting with (0,0), this will generate 1, 2; 2, 1; 0, 1; 1, 0; 2, 2; 0, 2; 1, 1; 2, 0. Thus all pairs. Similarly, the first and third positions in the generators are (1,2) and (0,2). Starting with (0,0), these generate 1, 2; 2, 1, 0, 2, 1, 1, 2, 0; 0, 1; 1, 0; 2, 2. These pairs of transformations work well because they are linearly independent and generate all the possible values. Note that if some pair were (1,1) and (2,2) for example, this wouldn't work. We would get 0, 0; 1, 1; 2, 2; and then repeat 2, 2 when applying the second transformation. So, we would not generate all possible combinations.

APPLICATION NOTE.

This technique is inspired by a clever approach to software testing pioneered by David M. Cohen, Siddhartha R. Dalal, Michael L. Fredman, and Gardner C. Patton of Telcordia. Their idea was to consider software to be a black box having discrete inputs (in our case, 10 inputs). Most errors, they assert, will be revealed when some combination of values (switch settings, in our case) of two of these inputs is tried. Apparently, this works well in practice.

HARD TO SURPRISE **SOLUTION**

1. Find surprising sequences that are as long as possible and that can be composed from 5, 10, or 26 distinct symbols.

The longest 2-surprising sequence composed from the first five letters of the alphabet is 12 characters long. Here is an example: For five letters (A .. E): AABCDECBDBAE. A long 2-surprising sequence from the first 10 letters of the alphabet is: BDGCJHHFACAEJG-FIEDIBHCGBFJ. And, a long 2-surprising sequence from all 26 letters of the alphabet, due to James J. Weinkam, is:

ABCDEFGHIJKLMNOPQRSTUVWXYZZYXWVUT-SRQPONACAFBLJMSIOHDGEKXGDHBFQCERJNPL.

2. What is the longest 3-surprising sequence you can find composed of the first five letters of the alphabet?

For a 3-surprising sequence built from five symbols, here is a long solution due to Andrew Palfreyman:
AABACDEBEDCCABDAEBC.

3. Can you find a theory to give the longest possible k-surprising sequences composed from sets of n symbols for any k and n.

One theoretical development first due to David Joyce and available at http://aleph0.clarku.edu/~djoyce/mpst/surprising/ is that no 2-surprising string on n characters can be longer than $3n$.

NOTE TO THE COMPUTATIONALLY MINDED.

One method of constructing long 2-surprising sequences for symbol alphabets of size k is to start with k strings each having one letter. As the algorithm proceeds, one has a set of strings that are 2-surprising of various lengths. Extend a 2-surprising string of length L with every possible symbol and then test whether the result is 2-surprising. If so, add that to the collection of strings of the length L + 1. To check whether a string is 2-surprising, you need only check whether the last symbol added violated a surprise. So if the last symbol is X, then look for other X's and the symbols preceding those X's to see whether any letter Y is the same distance D from a previous X as Y is from the last X.

The trouble with this method is that it is slow even for a computer when there are many symbols. Readers may find better methods. Note that if all we want to find is a sequence of maximum length, then instead of starting with k strings having one symbol, we can start with one string having one symbol and it doesn't matter which one. The reason is that any string of a given length L can be transformed to another one of the same length by a consistent one-to-one renaming. So there could be a different renaming for every first letter. More generally, if n letters have not yet been used at some string length, then instead of proceeding with n continuations, you can proceed with just one of those.

TANKTOPS AND SUNGLASSES
SOLUTION

1. What is the largest number of girls who could be in this gang and what might each girl wear in that case?

Here is a big gang of 12. Twelve is the largest number I know.

blue tanktop, black sunglass rims, black capris, pink lipstick;
blue tanktop, black sunglass rims, red capris, red lipstick;

blue tanktop, black sunglass rims, white capris, brown lipstick;

blue tanktop, brown sunglass rims, black capris, red lipstick;

blue tanktop, brown sunglass rims, red capris, pink lipstick;

black tanktop, black sunglass rims, black capris, red lipstick;

black tanktop, black sunglass rims, red capris, pink lipstick;

black tanktop, brown sunglass rims, black capris, pink lipstick;

black tanktop, brown sunglass rims, red capris, red lipstick;

blue tanktop, brown sunglass rims, pink capris, brown lipstick;

black tanktop, black sunglass rims, pink capris, brown lipstick;

black tanktop, brown sunglass rims, white capris, brown lipstick.

2. *What is the minimum number of girls who could be in this gang to satisfy the difference constraint but such that adding one girl would violate the constraint? Again show what each girl might wear in that case.*

Here there are only eight girls and none can be added. This would happen if the girls felt exclusionary.

blue tanktop, black sunglass rims, black capris, pink lipstick;

blue tanktop, brown sunglass rims, red capris, red lipstick;

black tanktop, black sunglass rims, red capris, brown lipstick;

black tanktop, brown sunglass rims, white capris, pink lipstick;

blue tanktop, brown sunglass rims, pink capris, brown lipstick;

black tanktop, black sunglass rims, pink capris, red lipstick;

blue tanktop, black sunglass rims, white capris, red lipstick;

black tanktop, brown sunglass rims, black capris, red lipstick.

It's OK. They grow out of it.

ABOUT THE AUTHOR

Dennis E. Shasha is a professor at New York University's Courant Institute, where he does research on biological pattern discovery, combinatorial pattern matching on trees and graphs, and the fast analysis of time series. After graduating from Yale in 1977, he worked for IBM designing circuits and microcode for the 3090 and completed his Ph.D. at Harvard in 1984. He is the author of several books, including three about a mathematical detective named Dr. Ecco entitled *The Puzzling Adventures of Dr. Ecco* (1988), *Codes, Puzzles, and Conspiracy* (1992), and *Dr. Ecco's Cyberpuzzles* (2002) and a book of biographies about great computer scientists called *Out of Their Minds: the Lives and Discoveries of 15 Great Computer Scientists* (1995). He currently writes monthly puzzle columns for *Scientific American* and *Dr. Dobb's Journal*.